Alexander Grushow
Rider University

STUDENT SOLUTIONS MANUAL

to accompany

CHEMISTRY

Structure and Dynamics

Fifth Edition

James Spencer
Franklin & Marshall College

George Bodner
Purdue University

Lyman Rickard
Millersville University

WILEY

John Wiley & Sons, Inc.

Cover Photo: © Aleksander Trankov/iStockphoto

To order books or for customer service, please call 1-800-CALL-WILEY (225-5945).

ISBN-13 978- 0-470-58712-6

10 9 8 7 6 5 4 3 2 1

Printed and bound by EPAC, Inc.

Table of Contents

Chapter 1
Elements and Compounds

1-1   The science of **chemistry** seeks to understand the composition, structure and properties of substances and the reactions by which one substance is converted into another. This is accomplished by performing experimental tests to recognize patterns of behavior among different substances, then developing models that explain these behaviors, and then using the models to predict behavior of other substances.

1-3   Experiments are an important part of chemistry. Through the observations made in an experiment, chemists develop a model to explain the results of the experiment. Further experiments are then performed to test the model's ability to predict the properties of other substances.

1-5   On the microscopic scale, an element consists of one particular type of atom. Nitrogen, N_2, oxygen, O_2, and iron metal, Fe, are examples of elements. A compound is made up of one molecule that contains more than one type of atom. Water, H_2O, and ammonia, NH_3, are examples of compounds. On a macroscopic scale, it is impossible to tell whether a sample is an element or a compound unless one tries to separate the sample into components. If it cannot be separated into different atoms, it is an element. If it can be separated, it is a compound.

1-7   Granite satisfies two criteria to consider it a mixture: it has variable composition and it can be separated.

1-9   The formula SO_3 indicates that the elements sulfur and oxygen are combined in the ratio of one to three. One atom of sulfur combines with three atoms of oxygen to form one molecular unit of SO_3.

1-11   (a)sodium   (b)magnesium   (c)aluminum   (d)silicon   (e)phosphorus   (f)chlorine   (g)argon

1-13   (a)molybdenum   (b)tungsten   (c)rhodium   (d)iridium   (e)palladium   (f)platinum   (g)silver   (h)gold   (i)mercury

1-15   The fact that chemical elements cannot be decomposed into simpler substances is consistent with the idea that there are elementary particles that can combine to form compounds but cannot be subdivided. The law of definite proportions argues strongly for the existence of atoms as a fundamental unit of combining. The seemingly continuous nature of matter as we sense it, the fluidity of the atmosphere, the fluidity of water, and the transparency of glass, all might be used to argue against the atomic nature of matter.

1-17   **Atomic theory** is widely accepted because the model has been successfully used to predict the outcomes of a multitude of experiments.

1-19   Atoms of different elements will have different weights and chemical properties.

1-21 $1 \text{ yr} \times \dfrac{365.25 \text{ day}}{1 \text{ yr}} \times \dfrac{24 \text{ hr}}{1 \text{ day}} \times \dfrac{3600 \text{ sec}}{1 \text{ day}} = 3.2 \times 10^7 \text{ sec}$

1-23 Since a centimeter has units of 10^{-2} meters, to convert from centimeters to the other units we need to apply the appropriate conversions. Note that since a centimeter is 10^{-2} meters, there are 100 centimeters in a meter. Similarly, since a micrometer is 10^{-6} meters, there are 10^6 micrometers in a meter.

$$4 \times 10^{-5} \text{ centimeters} \times \frac{1 \text{ meter}}{100 \text{ centimeter}} \times \frac{10^6 \text{ micrometers}}{1 \text{ meter}} = 4 \times 10^{-1} \text{ micrometers}$$

$$4 \times 10^{-5} \text{ centimeters} \times \frac{1 \text{ meter}}{100 \text{ centimeter}} \times \frac{10^9 \text{ nanometers}}{1 \text{ meter}} = 4 \times 10^2 \text{ nanometers}$$

1-25 $$100 \ \frac{ft^3}{min} \times \frac{1 \min}{60 \sec} \times \left(\frac{1 \text{ yard}}{3 \text{ ft}}\right)^3 \times \left(\frac{0.9144 \text{ m}}{1 \text{yard}}\right)^3 = 0.1416 \ \frac{m^3}{\sec}$$

1-27 (a) 3   The leading zeros (to the left) are not significant
(b) 4   The leading zeros (to the left) are not significant, but the trailing zero (to the right) is significant when a decimal point is shown
(c) 1   Trailing zeros are not significant when a decimal point is not present
(d) 5 The captured zeros (in between) are significant

1-29 (a) 475   (b) 0.0680   (c) 9.46×10^{10} (d) 30.1

1-31 (a) The first number has the fewest (2) digits to the right of the decimal point so we keep only those two digits upon addition: 153.92
(b) The first number has the fewest (none) digits to the right of the decimal point so we do not keep any digits to the right of the decimal point, but we still need to round appropriately after the addition: 33
(c) In multiplication we use the same number of figures as the number with the fewest total figures. Both numbers have three: 5.10×10^7
(d) Writing both in decimal form gives us 0.0418 + 0.00129. Using the rules above we should only keep four digits past the decimal: 0.0431

1-33 Dalton's assumption that all atoms of the same element are identical is in error. For example, it is possible for atoms of the same element to have a different number of neutrons.

1-35 Electron, relative charge: –1
Proton, relative charge: +1
Neutron, relative charge: 0

1-37 The electron is the particle that has the smallest mass.

1-39 The radius of an atom is approximately 10,000 times larger than the radius of the nucleus.

1-41 24 protons = chromium (Cr) = 24 atomic number
24 protons + 28 neutrons = 52 mass number
24 protons - 24 electrons = 0 net charge (a neutral atom)
^{52}Cr

1-43   Iodine (I) = 53 protons = 53 atomic number
       127 mass number - 53 protons = 74 neutrons
       neutral atom so the number of protons and electrons is equal: 53 electrons
       Z=53
       mass number = 127

1-45   34 protons=Selenium (Se)
       34 protons+45 neutrons=79 mass number
       34 protons-34 electrons=0 net charge; neutral atom
       ^{79}Se

1-47

	Z	A	e
^{31}P	15	31	15
^{18}O	8	18	8
^{39}K	19	39	19
^{58}Ni	28	58	28

1-49   The mass of ^{6}Li is 6.01512 amu. The mass of ^{1}H is 1.007825 amu. A ^{6}Li atom is
       5.96842 times more massive than a ^{1}H atom.

1-51 $\dfrac{\text{mass of } ^{12}\text{C}}{\text{mass of X}} = 0.750239$, mass of X =15.9949. Atom X is ^{16}O.

1-53

Mass (grams)	Z	A	Number of neutrons	Mass (amu)
1.6627×10^{-23}	5	10	5	10.0129
3.9829×10^{-23}	12	24	12	23.9850
2.9888×10^{-23}	8	18	10	17.9992
1.7752×10^{-22}	47	107	60	106.903

1-55   Three, ^{16}O, ^{17}O, ^{18}O

1-57 1.99265×10^{-23} g; 12.0000 amu

1-59   (b) The random selection will include isotopes ^{12}C and ^{13}C, but it will be mostly ^{12}C.

1-61   H$^+$ represents a hydrogen atom that has lost an electron to become a positive ion. The
       single electron of a hydrogen atom moves around the nucleus and makes the apparent size
       of the atom about 2000 times larger than the H$^+$ ion. H$_2$ represents a diatomic molecule in
       which two hydrogen atoms are chemically bonded to each other.

1-63   Atomic number = number of protons = 24, chromium (Cr)
       21 electrons which is three less than the number of protons, +3 charge
       mass number = 24 protons + 28 neutrons = 52
       ^{52}Cr^{+3}

1-65   34 protons=selenium (Se)
       34 protons+45 neutrons=79 mass number
       34 protons-36 electrons=-2 net charge
       $^{79}Se^{-2}$

1-67   Polyatomic ions are electronically charged substances that are composed of more than one atom. Note that a polyatomic ion is a single unit with a singular charge.

1-69   +1: Ammonium, NH_4^+
            Hydronium, H_3O^+

1-71   Compounds of the elements Cu and Ag with oxygen and chlorine, have the same formula as similar compounds of Li and Na. Ag_2O, Cu_2O, Li_2O Na_2O, AgCl, CuCl, LiCl, NaCl.

1-73   (a) IA   (b) IVA   (c) IIA   (d) VIA   (e) IIA   (f) VIIIA   (g) VIIA

1-75   Seven: H, Li, Na, K, Rb, Cs, Fr

1-77   Similar chemical properties are exhibited by elements in the same group. Sets which are in the same group are (a) and (d).

1-79   Gold is symbolized in the atomic world by its atomic symbol, Au.

1-81   The atomic weight of bromine is 0.5069 x 78.9183 amu + 0.4931 x 80.9163 amu = 79.90 amu.

1-83   At random, the total mass is 1.2011×10^6 amu. The average mass of a carbon atom is 12.011 amu. No single carbon atom has this mass.

1-85   126.90 amu

1-87   Since the atomic mass of an element is proportional to the relative percentages of its isotopes, this element would have a mass close to 11 amu. Boron with an atomic mass of 10.811 fits this criterion. The heavier isotope of boron would have 6 neutrons, 5 protons, and 5 electrons; the lighter would consist of 5 neutrons, 5 protons, and 5 electrons.

1-89   (a) 0.5184 x 106.90509 + 0.4816 x 108.90476 = 107.87 amu. This answer is the atomic mass of Ag given on the periodic table.
       (b) Both isotopes have 47 protons.
       (c) Both isotopes have 47 electrons. ^{107}Ag has 60 neutrons and ^{109}Ag has 62 neutrons.

1-91

	Z	A	e	n	%A
^6Li	3	6	3	3	7.42
^7Li	3	7	3	4	92.58
^{20}Ne	10	20	10	10	90.51
^{21}Ne	10	21	10	11	0.27
^{22}Ne	10	22	10	12	9.22

1-93   The average atomic mass cannot be greater than the largest mass isotope or smaller than the lowest mass isotope. Furthermore, the average atomic masses are listed on the periodic table.

1-95   The mass of the container and the contents <u>will not change</u>. The same number of atoms of each element will be present after burning the candle. However, the elements will be combined differently as product molecules.

1-97   Mass is conserved because atoms are not created or destroyed in a chemical reaction. Rather their arrangement changes to form different chemical compounds.

1-99   All the atoms that start out on the reactant side of a chemical reaction equation must be accounted for on the product side of the reaction equation.

1-101 A solid molecule of potassium iodide can react to form an aqueous plus one potassium ion and an aqueous minus one iodide ion.

1-103 (a) $4 Cr(s) + 3 O_2(g) \rightarrow 2 Cr_2O_3(s)$
       (b) $SiH_4(g) \rightarrow Si(s) + 2 H_2(g)$
       (c) $2 SO_3(g) \rightarrow 2 SO_2(g) + O_2(g)$

1-105 (a) $CH_4(g) + 2 O_2(g) \rightarrow CO_2(g) + 2 H_2O(g)$
       (b) $2 H_2S(g) + 3 O_2(g) \rightarrow 2 H_2O(g) + 2 SO_2(g)$
       (c) $2 B_5H_9(g) + 12 O_2(g) \rightarrow 5 B_2O_3(s) + 9 H_2O(g)$

1-107 (a) $1 C_3H_8(g) + 5 O_2(g) \rightarrow 3 CO_2(g) + 4 H_2O(g)$
       (b) $C_2H_5OH(l) + 3 O_2(g) \rightarrow 2 CO_2(g) + 3 H_2O(g)$
       (c) $C_6H_{12}O_6(s) + 6 O_2(g) \rightarrow 6 CO_2(g) + 6 H_2O(l)$

1-109   The ion has 18 electrons, 20 protons, and 20 neutrons. The chemical symbol for X is Ca.

1-111

classification	group	period	electrons	element
metal	IA	3	11	Na
semimetal	IVA	4	32	Ge
semimetal	IIIA	2	5	B
semimetal	IVA	3	14	Si
nonmetal	VIIA	4	35	Br

Chapter 2
The Mole: The Link Between the Macroscopic and the Atomic World of Chemistry

2-1 If ^{12}C = 1 amu then the mass of neon = $\dfrac{20.180 \text{ amu}}{12}$ = 1.6817 amu

2-3 (a) atomic weight of 6.941 amu and molar mass of 6.941 g
    (b) atomic weight of 12.011 amu and molar mass of 12.011 g
    (c) atomic weight of 24.305 amu and molar mass of 24.305 g
    (d) atomic weight of 63.546 amu and molar mass of 63.546g

2-5 1000 Si atoms, since the atomic weight is larger.

2-7 (a) 40.078 g
    (b) 87.62 g
    (c) 78.96 g
    (d) 72.61 g

2-9 A mole of atoms has a mass in grams equal to the atomic mass. Thus, the mass of one mole of chromium atoms is 51.996 g.

2-11 (a) 12.011 g C   (b) 58.693 g Ni   (c) 200.59 g Hg

2-13 Find the number of ^{12}C atoms in 12 lb of ^{12}C

$$\left(\frac{12 \text{ lb } ^{12}C}{\text{new mol}}\right)\left(\frac{454 \text{ g}}{\text{lb}}\right)\left(\frac{1 \text{ mol}}{12.000 \text{ g } ^{12}C}\right)\left(\frac{6.022 \times 10^{23} \text{ atoms}}{\text{mol}}\right) = \frac{2.7 \times 10^{26} \text{ atoms}}{\text{new mol}}$$

2-15 $\dfrac{1.762 \times 10^{-22} \text{ g}}{\text{molecule}} \times \dfrac{6.022 \times 10^{23} \text{ molecules}}{1 \text{ mol}} = \dfrac{106.1 \text{ g}}{\text{mol}}$ = molar mass of benzaldehyde

2-17 mass = 4.35×10^6 atoms $\times \dfrac{1 \text{ mol atoms}}{6.022 \times 10^{23} \text{ atoms}} \times \dfrac{12.000 \text{ g}}{\text{mol atoms}}$ = 8.67×10^{-17} g

2-19 $^{12}C^{16}O_2$ = 1 x 12.000amu + 2 x 16.000 amu = 44.000 amu x $\left(\dfrac{1 \text{ g}}{6.022 \times 10^{23} \text{amu}}\right)$

    = 7.306×10^{-23} g.

2-21 $1.65 \text{ mol Cu} \times \dfrac{63.546 \text{ g}}{1 \text{ mol Cu}} = 105 \text{ g}$

2-23 $2 \text{ mol atoms} \times \dfrac{6.022 \times 10^{23} \text{ atoms}}{1 \text{ mol atoms}} = 1.2044 \times 10^{24} \text{ atoms}$

2-25 $\dfrac{55.847 \text{ amu}}{\text{atom}} \times \dfrac{1 \text{ g}}{6.022 \times 10^{23} \text{ amu}} = 9.2738 \times 10^{-23} \dfrac{\text{g}}{\text{atom}}$

2-27   12 methane molecules contain 12 carbon atoms,12 methane molecules contain 4 x 12 = 48 hydrogen atoms. One mole of methane molecules contains one mole, 6.022 x 10^{23}, carbon atoms. One mole of methane molecules, 6.022 x 10^{23} molecules, contains 4 x 6.022 x 10^{23} = 24.09 x 10^{24} hydrogen atoms.

2-29 $$\left(6.02 \times 10^{23} \text{ molecules } H_2\right) \left(\frac{2 \text{ H atoms}}{\text{molecule } H_2}\right) = 1.20 \times 10^{24} \text{ atoms H}$$

$$\left(1.00 \text{ mol } H_2\right) \left(\frac{6.022 \times 10^{23} \text{ molecules}}{\text{mol}}\right) = 6.02 \times 10^{23} \text{ molecules } H_2$$

$$\left(1.00 \text{ mol } H_2\right) \left(\frac{2.0158 \text{ g } H_2}{\text{mol } H_2}\right) = 2.02 \text{ g } H_2$$

2-31   (d)   4 mol NH_3 and 3 mol N_2H_4 both have 12 hydrogen atoms.

2-33 HCO_2H, 2(1.0079)+12.011+2(15.999)=46.025 $\frac{g}{mol}$

H_2CO, 2(1.0079)+12.011+15.999 = 30.026 $\frac{g}{mol}$

2-35   (a) P_4S_{10}, 4(30.974) + 10(32.066) = 444.556 $\frac{g}{mol}$

(b) NO_2, 14.007 + 2(15.999) = 46.005 $\frac{g}{mol}$

(c) ZnS, 65.39 + 32.066 = 97.46 $\frac{g}{mol}$

(d) $KMnO_4$, 39.098 + 54.938 + 4(15.999) = 158.032 $\frac{g}{mol}$

2-37   Molar mass of safrole, $C_{10}H_{10}O_2$:

10(12.011) + 10(1.0079) + 2(15.999) = 162.187 $\frac{g}{mol}$

2-39   (a)   Darvon, $C_{22}H_{30}ClNO_2$:

22(12.011) + 30(1.0079) + 35.453 + 14.007 + 2(15.999) = 375.937 $\frac{g}{mol}$

(b)   Valium, $C_{16}H_{13}ClN_2O$:

16(12.011) + 13(1.0079) + 35.453 + 2(14.007) + 15.999 = 284.745 $\frac{g}{mol}$

(c)   Tetracycline, $C_{22}H_{24}N_2O_8$:

22(12.011) + 24(1.0079) + 2(14.007) + 8(15.999) = 444.438 $\frac{g}{mol}$

2-41 $$\left(0.0582 \text{ mol } CCl_4\right) \left(\frac{153.823 \text{ g } CCl_4}{1 \text{ mol } CCl_4}\right) = 8.95 \text{ g } CCl_4$$

2-43 $$\left(5.72 \text{ g Al}\right) \left(\frac{1 \text{ mol Al}}{26.982 \text{ g Al}}\right) = 0.212 \text{ mol Al}$$

2-45 (a) $(NH_4)_2SO_4$, $\%N = \dfrac{2(14.007)}{132.139} \times 100 = 21.200\ \%$

(b) KNO_3, $\%N = \dfrac{14.007}{101.102} \times 100 = 13.854\ \%$

(c) $NaNO_3$, $\%N = \dfrac{14.007}{84.994} \times 100 = 16.480\ \%$

(d) $(H_2N)_2CO$, $\%N = \dfrac{2(14.007)}{60.056} \times 100 = 46.646\ \%$

2-47 Molar mass of $Be_3Al_2(SiO_3)_6$:

$3(9.0122) + 2(26.982) + 6(28.086) + 18(15.999) = 537.499\ \dfrac{g}{mol}$

$\%Si = \dfrac{6(28.086)}{537.499} \times 100 = 31.352\ \%\ Si$

2-49 $(0.244\ g\ CaC_2)\left(\dfrac{1\ mol\ CaC_2}{64.100\ g\ CaC_2}\right)\left(\dfrac{2\ mol\ C}{1\ mol\ CaC_2}\right) = 7.61 \times 10^{-3}\ mols\ C$

2-51 $K_2PtCl_6 = 2(39.098) + 195.08 + 6(35.453) = 486.00\ amu$

$(0.756\ g\ K_2PtCl_6)\left(\dfrac{1\ mol\ K_2PtCl_6}{485.99\ g\ K_2PtCl_6}\right)\left(\dfrac{6\ mol\ Cl}{1\ mol\ K_2PtCl_6}\right)\left(\dfrac{6.022\times 10^{23}\ atoms\ Cl}{1\ mol\ Cl}\right)$

$= 5.62 \times 10^{21}\ atoms\ of\ Cl$

2-53 (a) A 100.0 g sample contains 36.8 g N and 63.2 g O.

(b) $(36.8\ g\ N)\left(\dfrac{1\ mol\ N}{14.007\ g\ N}\right) = 2.63\ mols\ N$

$(63.2\ g\ O)\left(\dfrac{1\ mol\ O}{15.999\ g\ O}\right) = 3.95\ mols\ O$

(c) $\dfrac{3.95}{2.63}$ or $1.50:1$

(d) N_2O_3

2-55 A 100.00 g sample of magnetite contains 72.36 g Fe and 27.64 g O.

mol Fe $= 72.36\ g \times \dfrac{1\ mol\ Fe}{55.847\ g} = 1.296\ mol\ Fe$

mol O $= 27.64\ g \times \dfrac{1\ mol\ O}{15.999\ g} = 1.728\ mol\ O$

to determine the ratio $\dfrac{1.296}{1.296} = 1.000 = 1\ mol\ Fe$ $\quad \dfrac{1.728}{1.296} = 1.334 = 1.334\ mol\ O$

$Fe_1O_{1.334}$

Since an empirical formula contains the simplest whole number ratio of combination, the formula of magnetite is Fe_3O_4.

2-57 A 100.00 g sample of nitrous oxide contains 63.65 g N and 36.35 g O.

mol N $= 63.65\ g\ N \times \dfrac{1\ mol\ N}{14.007\ g} = 4.544\ mol\ N$ $\quad \dfrac{4.544}{2.272} = 2.000 = 2\ mol\ N$

mol O $= 36.35\ g\ O \times \dfrac{1\ mol\ O}{15.999\ g} = 2.272\ mol\ O$ $\quad \dfrac{2.272}{2.272} = 1.000 = 1\ mol\ O$

Nitrous oxide has the formula N_2O.

2-59   A 100.0 g sample 53.5 g Xe and 100.0-53.5 =46.5 g F.

$\text{\# mol Xe} = 53.5 \text{ g Xe} \dfrac{1 \text{ mol Xe}}{131.29 \text{ g}} = 0.407 \text{ mol Xe}$ $\dfrac{0.407}{0.407} = 1.00 = 1 \text{ mol Xe}$

$\text{\# mol F} = 46.5 \text{ g F} \times \dfrac{1 \text{ mol F}}{18.998 \text{ g}} = 2.45 \text{ mol F}$ $\dfrac{2.45}{0.407} = 6.02 = 6 \text{ mol F}$

The compound has the empirical formula XeF_6.

2-61 $\text{\# mol N} = 0.483 \text{ g} \times \dfrac{1 \text{ mol N}}{14.007 \text{ g}} = 3.45 \times 10^{-2} \text{ mol N}$ $\dfrac{3.45 \times 10^{-2}}{3.45 \times 10^{-2}} = 1 \text{ mol N}$

$\text{\# mol O} = 1.104 \text{ g} \times \dfrac{1 \text{ mol O}}{15.999 \text{ g}} = 6.900 \times 10^{-2} \text{ mol O}$ $\dfrac{6.900 \times 10^{-2}}{3.45 \times 10^{-2}} = 2 \text{ mol O}$

The empirical formula is NO_2; response (c).

2-63   No, only the empirical formula can be determined from percent composition. The molar mass must be known to determine the molecular formula.

2-65   A 100.00 g sample of phenolphthalein contains 75.46 g C, 4.43 g H, and 20.10 g O.

$\text{\# mol C} = 75.46 \text{ g} \times \dfrac{1 \text{ mol C}}{12.011 \text{ g}} = 6.283 \text{ mol C}$ $\dfrac{6.283}{1.256} = 5.000 = 5 \text{ mol C}$

$\text{\# mol H} = 4.43 \text{ g} \times \dfrac{1 \text{ mol H}}{1.0079 \text{ g}} = 4.40 \text{ mol H}$ $\dfrac{4.40}{1.256} = 3.50 = 3.5 \text{ mol H}$

$\text{\# mol O} = 20.10 \text{ g O} \times \dfrac{1 \text{ mol O}}{15.999 \text{ g}} = 1.256 \text{ mol O}$ $\dfrac{1.256}{1.256} = 1.000 = 1 \text{ mol O}$

To get a whole number ratio, we must multiply by 2.
The empirical formula is $C_{10}H_7O_2$ which has a molar mass of 159.163. Since the molar mass is 318.327 g/mol, the molecular formula is $C_{20}H_{14}O_4$.

2-67   A 100.00 g sample of aspartame contains 57.14 g C, 6.16 g H, 9.52 g N, and 27.18 g O.

$\text{\# mol C} = 57.14 \text{ g C} \times \dfrac{1 \text{ mol C}}{12.011 \text{ g}} = 4.757 \text{ mol C}$ $\dfrac{4.757}{0.680} = 7.00 = 7 \text{ mol C}$

$\text{\# mol H} = 6.16 \text{ g H} \times \dfrac{1 \text{ mol H}}{1.0079 \text{ g}} = 6.11 \text{ mol H}$ $\dfrac{6.11}{0.680} = 8.99 = 9 \text{ mol H}$

$\text{\# mol N} = 9.52 \text{ g N} \times \dfrac{1 \text{ mol N}}{14.007 \text{ g}} = 0.680 \text{ mol N}$ $\dfrac{0.680}{0.680} = 1.00 = 1 \text{ mol N}$

$\text{\# mol O} = 27.18 \text{ g O} \times \dfrac{1 \text{ mol O}}{15.999 \text{ g}} = 1.699 \text{ mol O}$ $\dfrac{1.699}{0.680} = 2.50 = 2.5 \text{ mol O}$

To get a whole number ratio, we must multiply by 2. The empirical formula of aspartame is $C_{14}H_{18}N_2O_5$ and has a mass of 294.305 g/mol. Since the molar mass is also 294.305 g/mol, the molecular formula and the empirical formula are the same.

2-69   Reactants: 3 Ca (3x40.078 amu) + N_2 (28.014 amu) = 134.248 amu
Products: Ca_3N_2 (3x40.078 amu+2x14.007)=134.248 amu
When using mole quantities the amu unit can be replaced with a grams unit and the numbers remain the same.

2-71   From 5 mol O_2 x $\dfrac{2 \text{ mol CO}_2}{1 \text{ mol O}_2}$ = 10 mol CO_2

2-73   12 mol Cu x $\dfrac{1 \text{ mol CuO}}{1 \text{ mol Cu}}$ = 12 mol CuO

2-75 $3 MnO_2(s) \rightarrow Mn_3O_4(s) + O_2(g)$

$$6.75 \text{ mol MnO} \times \frac{1 \text{ mol } O_2}{3 \text{ mol } MnO_2} = 2.25 \text{ mol } O_2$$

2-77 1. Convert grams of CO reacted to moles of CO.
2. Relate moles of CO reacted to moles of CO_2 produced.
3. Convert moles CO_2 to grams.

2-79 $CH_4(g) + 2 O_2(g) \rightarrow CO_2(g) + 2 H_2O(g)$

$$10.0 \text{ g } CH_4 \times \frac{1 \text{ mol } CH_4}{16.043 \text{ g}} \times \frac{2 \text{ mol } O_2}{1 \text{ mol } CH_4} \times \frac{31.998 \text{ g } O_2}{1 \text{ mol } O_2} = 39.9 \text{ g } O_2 \text{ consumed.}$$

$$10.0 \text{ g } CH_4 \times \frac{1 \text{ mol } CH_4}{16.043 \text{ g}} \times \frac{1 \text{ mol } CO_2}{1 \text{ mol } CH_4} \times \frac{44.009 \text{ g } CO_2}{1 \text{ mol } CO_2} = 27.4 \text{ g } CO_2 \text{ produced.}$$

2-81 $2 KClO_3(s) \rightarrow 2 KCl(s) + 3 O_2(s)$

$$25.0 \text{ g } KClO_3 \times \frac{1 \text{ mol } KClO_3}{122.548 \text{ g}} \times \frac{3 \text{ mol } O_2}{2 \text{ mol } KClO_3} \times \frac{31.998 \text{ g } O_2}{1 \text{ mol } O_2} = 9.79 \text{ g } O_2$$

2-83 $C_6H_{12}O_6(aq) \rightarrow 2 C_2H_5OH(aq) + 2 CO_2(g)$

$$1.00 \text{ kg } C_6H_{12}O_6 \times \frac{1000 \text{ g}}{1 \text{ kg}} \times \frac{1 \text{ mol } C_6H_{12}O_6}{180.155 \text{ g}} \times \frac{2 \text{ mol } C_2H_5OH}{1 \text{ mol } C_6H_{12}O_6} \times \frac{46.068 \text{ g } C_2H_5OH}{1 \text{ mol } C_2H_5OH} \times \frac{1 \text{ kg}}{1000 \text{ g}}$$

$$= 0.511 \text{ kg } C_2H_5OH$$

2-85 $Ca_3P_2(s) + 6 H_2O(l) \rightarrow 3 Ca(OH)_2(aq) + 2 PH_3(g)$

$$10.0 \text{ g } Ca_3P_2 \times \frac{1 \text{ mol } Ca_3P_2}{182.182 \text{ g}} \times \frac{2 \text{ mol } PH_3}{1 \text{ mol } Ca_3P_2} \times \frac{33.998 \text{ g } PH_3}{1 \text{ mol } PH_3} = 3.73 \text{ g } PH_3$$

2-87 $N_2(g) + 3 H_2(g) \rightarrow 2 NH_3(g)$
$4 NH_3(g) + 5 O_2(g) \rightarrow 4 NO(g) + 6 H_2O(g)$
$2 NO(g) + O_2 \rightarrow 2 NO_2(g)$
$3 NO_2(g) + H_2O(l) \rightarrow 2 HNO_3(aq) + NO(g)$

$$150 \text{ g } HNO_3 \times \frac{1 \text{ mol } HNO_3}{63.012 \text{ g}} \times \frac{3 \text{ mol } NO_2}{2 \text{ mol } HNO_3} \times \frac{2 \text{ mol } NO}{2 \text{ mol } NO_2} \times \frac{4 \text{ mol } NH_3}{4 \text{ mol } NO} \times \frac{1 \text{ mol } N_2}{2 \text{ mol } NH_3} \times$$

$$\frac{28.014 \text{ g } N_2}{1 \text{ mol } N_2} = 50.0 \text{ g } N_2$$

2-89 $4 P_4(s) + 5 S_8(s) \rightarrow 4 P_4S_{10}(s)$

$$0.500 \text{ mol } P_4 \times \frac{4 \text{ mol } P_4S_{10}}{4 \text{ mol } P_4} = 0.500 \text{ mol } P_4S_{10}$$

$$0.500 \text{ } S_8 \times \frac{4 \text{ mol } P_4S_{10}}{5 \text{ mol } S_8} = 0.400 \text{ mol } P_4S_{10}$$

Since S_8 produces fewer moles of P_4S_{10}, it is the limiting reagent. If P_4 is doubled, S_8 is still the limiting reagent, and the amount of P_4S_{10} produced would remain unchanged. If S_8 is doubled, then P_4 becomes the limiting reagent and the yield of P_4S_{10} would be 0.500 mol.

2-91 $H_2(g) + Cl_2(g) \rightarrow 2\ HCl(g)$

$$10.0\ g\ H_2 \times \frac{1\ mol\ H_2}{2.0158\ g} \times \frac{2\ mol\ HCl}{1\ mol\ H_2} = 9.92\ mol\ HCl$$

$$10.0\ g\ Cl_2 \times \frac{1\ mol\ Cl_2}{70.906\ g} \times \frac{2\ mol\ HCl}{1\ mol\ Cl_2} = 0.282\ mol\ HCl$$

Cl_2 produces fewer moles of HCl, therefore it is the limiting reagent.

$$0.282\ mol\ HCl \times \frac{36.46g}{mol} = 10.3\ g\ HCl$$

To increase the amount of HCl produced, the amount of Cl_2 would have to be increased.

2-93 $2\ PF_3(g) + XeF_4(s) \rightarrow 2\ PF_5(g) + Xe(g)$

$$100.0\ g\ PF_3 \times \frac{1\ mol\ PF_3}{87.968\ g} \times \frac{2\ mol\ PF_5}{2\ mol\ PF_3} = 1.137\ mol\ PF_5$$

$$50.0\ g\ XeF_4 \times \frac{1\ mol\ XeF_4}{207.28\ g} \times \frac{2\ mol\ PF_5}{1\ mol\ XeF_4} = 0.482\ mol\ PF_5$$

XeF_4 produces fewer moles of PF_5; therefore, it is the limiting reagent and 0.482 moles of PF_5 would be produced.

2-95 $Fe_2O_3(s) + 2\ Al(s) \rightarrow Al_2O_3(s) + 2\ Fe(l)$

$$150\ g\ Al \times \frac{1\ mol\ Al}{26.982\ g} \times \frac{2\ mol\ Fe}{2\ mol\ Al} = 5.56\ mol\ Fe,$$

$$250\ g\ Fe_2O_3 \times \frac{1\ mol\ Fe_2O_3}{159.691g\ Fe_2O_3} \times \frac{2\ mol\ Fe}{1\ mol\ Fe_2O_3} = 3.13\ mol\ Fe,$$

Fe_2O_3 produces fewer moles of Fe; therefore it is the limiting reagent. The amount of Fe produced is

$$3.13\ mol\ Fe \times \frac{55.847\ g}{1\ mol\ Fe} = 175\ g\ Fe.$$

2-97 $\dfrac{100g}{8.8mL} = 11\ \dfrac{g}{cm}$: The first strip is Pb. $\qquad \dfrac{100g}{37.0mL} = 2.70\ \dfrac{g}{cm}$: The second strip is Al.

2-99 $5.6\ mL \times 13.6\ \dfrac{g}{cm} = 76\ g$

2-101 Both (b) and (c) are homogeneous mixtures.

2-103 $27.3\ g\ HCl \times \dfrac{1\ mol\ HCl}{36.461g\ HCl} = 0.749\ mol\ HCl$

$$M = \frac{moles}{liter} = \frac{0.749\ mol}{0.125\ L} = 5.99\ M$$

2-105 $252\ g\ NH_3 \times \dfrac{1\ mol\ NH_3}{17.031g} = 14.8\ mol\ NH_3$

$$M = \frac{moles}{L} = \frac{14.8\ mol}{1\ L} = 14.8\ M$$

2-107 $5.77\ g\ Cl_2 \times \dfrac{1\ mol\ Cl_2}{70.906\ g} = 0.0814\ mol\ Cl_2$

$$M = \frac{mol}{L} = \frac{0.0814\ mol}{1.00\ L} = 0.0814\ M$$

2-109 1.00 L of water plus 1.00 mol of K_2CrO_4 may not be 1.00 L of solution. The student probably made more than 1.00 L of solution, so the solution was less than 1.00 M. A 1.00 liter sample of a 1.00 mol solute per liter solution must be prepared by placing the solute in a container calibrated to hold 1.000 liter. Add water to the container (volumetric flask) to dissolve the solute. Thoroughly mix the solution. Continue to add more water (solvent) until the liquid level has been brought to the calibration mark.

2-111 $\dfrac{20\ ng}{ml} \times \dfrac{10^{-9}\ g}{ng} \times \dfrac{1000\ mL}{L} \times \dfrac{1\ mol}{315\ g} = 6.3 \times 10^{-8}\ M$

2-113 $\dfrac{2.75\ g\ AgNO_3}{0.250\ L} \times \dfrac{1\ mol\ AgNO_3}{169.87\ g\ AgNO_3} = 0.0648\ M\ AgNO_3$. To make it half as concentrated either add half as much $AgNO_3$ or use twice as much water.

2-115 $0.50\dfrac{mol}{L} \times 0.500\ L = 0.25\ mol$

2-117 The 0.25 M solution is more concentrated since it is the higher molarity; more moles per liter.

2-119 (a) $\dfrac{0.275\ g\ AgNO_3}{0.500\ L} \times \dfrac{1\ mol\ AgNO_3}{169.87\ g\ AgNO_3} = 0.00324\ M\ AgNO_3$

(b) $0.00324\ \dfrac{mol}{L}\ AgNO_3 \times \dfrac{0.0100\ L}{0.500\ L} = 6.48 \times 10^{-5}\ M\ AgNO_3$

(c) $6.48 \times 10^{-5}\ \dfrac{mol}{L}\ AgNO_3 \times \dfrac{0.0100\ L}{0.250\ L} = 2.60 \times 10^{-6}\ M\ AgNO_3$

2-121 $0.050\dfrac{mol}{L}\ CuSO_4 \times \dfrac{x\ L}{2 \cdot x\ L} = 0.025\ M\ CuSO_4$

2-123 $0.10\dfrac{mol}{L}\ HCl \times 0.250\ L \times \dfrac{1\ L}{6.0\ mol\ HCl} = 0.0042\ L$. 4.2 mL of 6.0 M HCl would be diluted to 250 mL. The resulting solution will be 0.10 M HCl.

2-125 $1.20\ \dfrac{mol}{L}\ KF \times 0.100\ L = 0.120\ mol\ KF$ are present in the initial solution. If you want a final concentration of 0.45 M, what volume must contain the 0.120 mol?

$0.120\ mol \times \dfrac{1\ L}{0.45\ mol} = .267\ L$ is the final volume that the 100 mL is diluted to.

2-127 $M_1V_1 = M_2V_2$
(1.00 L)(3.00 M)=(x L)(17.4 M)

$\dfrac{(1.00\ L)(3.00\ M)}{(17.4\ M)} = 0.172\ L = 172$ mL of 17.4 M acetic acid is needed.

2-129 $(0.200 \text{ L})(1.25 \text{ M})=(x \text{ L})(5.94 \text{ M})$

$\dfrac{(0.200 \text{ L})(1.25 \text{ M})}{5.94 \text{ M}} = x \text{ L} = 0.0421 \text{ L} = 42.1 \text{ mL}$

Take about 150 mL distilled water and slowly add 42.1 mL of the HNO_3 with mixing. Bring the solution to a final volume of 200 mL with distilled water and mix. The resulting solution is 1.25 M HNO_3.

2-131 $2 \text{ NaI(aq)} + \text{Hg(NO}_3)_2\text{(aq)} \rightarrow \text{HgI}_2\text{(s)} + \text{NaNO}_3\text{(aq)}$

$0.045 \text{ L} \times \dfrac{0.10 \text{ mol Hg(NO}_3)_2}{1 \text{ L}} = 4.5 \times 10^{-3} \text{ mol Hg(NO}_3)_2$

$4.5 \times 10^{-3} \text{ mol Hg(NO}_3)_2 \times \dfrac{2 \text{ mol NaI}}{1 \text{ mol Hg(NO}_3)_2} = 9.00 \times 10^{-3} \text{ mol NaI}$

$\dfrac{9.00 \times 10^{-3} \text{ mol NaI}}{0.25 \dfrac{\text{mol NaI}}{\text{L}}} = 3.6 \times 10^{-2} \text{ L} \times \dfrac{1000 \text{ mL}}{1 \text{ L}} = 36 \text{ mL NaI}$

2-133 $H_2C_2O_4\text{(aq)} + 2 \text{ NaOH(aq)} \rightarrow \text{Na}_2C_2O_4\text{(aq)} + 2 H_2O\text{(l)}$

$25.00 \text{ mL } H_2C_2O_4 \times \dfrac{0.2043 \text{ mol } H_2C_2O_4}{1000 \text{ mL}} = 5.108 \times 10^{-3} \text{ mol } H_2C_2O_4$

$5.108 \times 10^{-3} \text{ mol } H_2C_2O_4 \times \dfrac{2 \text{ mol NaOH}}{1 \text{ mol } H_2C_2O_4} = 1.022 \times 10^{-2} \text{ mol NaOH}$

$M = \dfrac{\text{mol}}{\text{L}} = \dfrac{1.022 \times 10^{-2} \text{ mol NaOH}}{0.01042 \text{ L}} = 0.9808 \text{ M NaOH}$

2-135 $C_6H_{12}O_6\text{(aq)} + 5 \text{ IO}_4^-\text{(aq)} \rightarrow 5 \text{ IO}_3^-\text{(aq)} + 5 \text{ HCO}_2\text{H(aq)} + H_2\text{CO(aq)}$

$25.0 \text{ mL IO}_4^- \times \dfrac{0.750 \text{ mol IO}_4^-}{1000 \text{ mL}} = 0.0188 \text{ mol IO}_4^-$

$0.0188 \text{ mol IO}_4^- \times \dfrac{1 \text{ mol } C_6H_{12}O_6}{5 \text{ mol IO}_4^-} = 3.75 \times 10^{-3} \text{ mol } C_6H_{12}O_6$

$M = \dfrac{\text{mol}}{\text{L}} = \dfrac{3.75 \times 10^{-3} \text{ mol } C_6H_{12}O_6}{0.0100 \text{ L}} = 0.375 \text{ M } C_6H_{12}O_6$

2-137   Mass of original sample: 5.00 g $BaCl_2 \cdot xH_2O$

Mass of sample after heating: 4.26 g $BaCl_2$

Mass of H_2O driven off: 5.00 g - 4.26 g = 0.74 g H_2O

Moles of H_2O driven off: $0.74 \text{ g } H_2O \times \dfrac{1 \text{ mol } H_2O}{18.015 \text{ g}} = 4.1 \times 10^{-2} \text{ mol } H_2O$

Moles of $BaCl_2$ remaining: $4.26 \text{ g } BaCl_2 \times \dfrac{1 \text{ mol } BaCl_2}{208.24 \text{ g}} = 2.05 \times 10^{-2} \text{ mol } BaCl_2$

Since there are twice as many moles of H_2O per mole of $BaCl_2$, the formula for the hydrate is $BaCl_2 \cdot 2H_2O$.

2-139 3.500 g compound -1.288 g O$=2.212$ g Mn

$$1.288 \text{ g O} \times \frac{1 \text{ mol O}}{15.999 \text{ g}} = 8.051 \times 10^{-2} \text{ mol O}$$

$$2.212 \text{ g Mn} \times \frac{1 \text{ mol Mn}}{54.938 \text{ g}} = 4.026 \times 10^{-2} \text{ mol Mn}$$

divide by smallest

$$\frac{8.051 \times 10^{-2} \text{ mol O}}{4.026 \times 10^{-2}} = 2 \text{ mol O} \qquad \frac{4.026 \times 10^{-2} \text{ mol Mn}}{4.026 \times 10^{-2}} = 1 \text{ mol Mn}$$

The compound has the empirical formula MnO_2.

2-141 % CO_2 from $CaCO_3 = 43.97\%$ (see problem 2-85)
CaO remaining is 56.03%
grams $CaCO_3 = 42.670 - 35.351 = 7.319$ g $CaCO_3$
grams CO_2 evolved$=(0.4397)(7.319 \text{ g})=3.218$ g CO_2
gram residue$=7.319-3.218=4.101$ g CaO residue
Theoretical mass crucible + residue$=35.351+4.101=39.452$ g

2-143 $CuS + 2 H_2SO_4(aq) \rightarrow CuSO_4(aq) + SO_2(g) + 2 H_2O(l)$
$2 CuSO_4(aq) + 5 I^-(aq) \rightarrow 2 CuI(s) + I_3^-(aq) + 2 SO_4^{2-}$
$I_3^-(aq) + 2 S_2O_3^{2-}(aq) \rightarrow 3 I^-(aq) + S_4O_6^{2-}(aq)$

$$31.5 \text{ mL} \times \frac{1.00 \text{ mol } S_2O_3^{2-}}{1000 \text{ ml}} = 0.0315 \text{ mol } S_2O_3^{2-}$$

$$\% \text{ Cu} = 0.0315 \text{ mol } S_2O_3^{2-} \times \frac{1 \text{ mol } I_3^-}{2 \text{ mol } S_2O_3^{2-}} \times \frac{2 \text{ mol } CuSO_4}{1 \text{ mol } I_3^-} \times \frac{1 \text{ mol Cu}}{1 \text{ mol } CuSO_4}$$

$$\times \frac{63.546 \text{ g Cu}}{1 \text{ mol Cu}} \times \frac{1}{2.50 \text{ g sample}} \times 100 = 80.0\%$$

2-145 $4 Fe(s) + 3 O_2(g) \rightarrow 2 Fe_2O_3(s)$
$3 Fe(s) + 2 O_2(g) \rightarrow Fe_3O_4(s)$

Moles of Fe reacted: $167.6 \text{ g Fe} \times \frac{1 \text{ mol Fe}}{55.847 \text{ g}} = 3.001 \text{ mol Fe}$

Grams of O in the iron oxide: $231.6 \text{ g} - 167.6 \text{ g} = 64.0$ g O

Moles of O in the iron oxide: $64.0 \text{ g O} \times \frac{1 \text{ mol O}}{15.999 \text{ g}} = 4.000 \text{ mol O}$

The oxide formed is Fe_3O_4.

Chapter 3
The Structure of the Atom

3-1 The positive charge is concentrated in a small portion of the atom called the nucleus.

3-3 According to the Rutherford model the nucleus contains the positive charge of the atom and contains most of the mass of the atom. At the time of Rutherford's experiments the distinction between protons and neutrons had not been made.

3-5 Particles have a definite mass and occupy space. Waves have no mass and carry energy as they travel through space. In addition waves have properties of speed, frequency, wavelength, and amplitude.

3-7 For 440 Hz: Assume the speed of sound is 1116 ft/sec

$$\left(\frac{440 \text{ cycles}}{s}\right) \times \lambda = 1116 \frac{ft}{s}$$

$$\lambda = \left(1116 \frac{ft}{s}\right)\left(\frac{1 \text{ s}}{440 \text{ cycles}}\right) = 2.54 \frac{ft}{cycle}$$

For 880 Hz

$$\left(\frac{880 \text{ cycles}}{s}\right) \times \lambda = 1116 \frac{ft}{s}$$

$$\lambda = \left(1116 \frac{ft}{s}\right)\left(\frac{1 \text{ s}}{880 \text{ cycles}}\right) = 1.27 \frac{ft}{cycle}$$

As the frequency increased by a factor of 2 the wavelength is decreased by a factor of 2. The speed at which the sound travels to your ear remains constant at 1116 ft/s.

3-9 $\nu\lambda=c$

$$\left(5.0 \times 10^{14} \frac{\text{cycles}}{s}\right) \times \lambda = 2.998 \times 10^8 \frac{m}{s}$$

$$\lambda = 2.998 \times 10^8 \frac{m}{s} \times \left(\frac{s}{5.0 \times 10^{14} \text{ cycles}}\right) = 6.0 \times 10^{-7} \frac{m}{cycle}$$

3-11 Red light has a longer wavelength than blue light.

3-13 $25.147 \text{ MHz} \times \left(\frac{1 \times 10^6 \text{ Hz}}{1 \text{ MHz}}\right) = 2.5147 \times 10^7 \text{ Hz}$

$\nu\lambda=c$

$$\left(2.5147 \times 10^7 \frac{\text{cycles}}{s}\right) \times \lambda = 2.998 \times 10^8 \frac{m}{s}$$

$$\lambda = \left(2.998 \times 10^8 \frac{m}{s}\right)\left(\frac{1 \text{ s}}{2.5147 \times 10^7 \text{ cycles}}\right) = 11.92 \frac{m}{cycle}$$

This radiation falls in the radio-wave region of the electromagnetic spectrum.

3-15 $\lambda_1 = 668 \text{ nm} = \text{red light}$

$\lambda_2 = 609 \text{ nm} = \text{orange light}$

3-17 $\lambda_1 = 588.9953$ nm $= 5.889953 \times 10^{-7}$ m

$$\nu_1 = 2.997924 \times 10^8 \ \frac{m}{s} \times \left(\frac{1 \ cycle}{5.889953 \times 10^{-7}m} \right) = 5.089894 \times 10^{14} \ \frac{cycle}{s}$$

$\lambda_2 = 589.5923$ nm $= 5.895923 \times 10^{-7}$ m

$$\nu_2 = 2.997924 \times 10^8 \ \frac{m}{s} \times \left(\frac{1 \ cycle}{5.895923 \times 10^{-7}m} \right) = 5.084740 \times 10^{14} \ \frac{cycle}{s}$$

3-19 The shorter the wavelength the greater the energy; 580 nm.

3-21 $\lambda = 656.3$ nm $= 6.563 \times 10^{-7}$ m

$$\nu = 2.998 \times 10^8 \ \frac{m}{s} \times \left(\frac{1 \ cycle}{6.563 \times 10^{-7} \ m} \right) = 4.568 \times 10^{14} \ \frac{cycle}{s}$$

$E = h\nu$

$$E = (6.626 \times 10^{-34} \ J \ s) \times \left(4.568 \times 10^{14} \ \frac{cycles}{s} \right) = 3.027 \times 10^{-19} \ J$$

3-23
$$\left(\frac{243.4 \ kJ}{mole} \right)\left(\frac{1000 J}{kJ} \right)\left(\frac{1 \ mole}{6.022 \times 10^{23} \ molecules \ Cl_2} \right) = \frac{4.042 \times 10^{-19} J}{molecule \ Cl_2}$$

$E = h\nu = hc/\lambda$

$\lambda = hc/E$

$$\lambda = \left(\frac{6.626 \times 10^{-34} J \ s \ \times \ 2.998 \times 10^8 m/s}{4.042 \times 10^{-19} J/molecule \ Cl_2} \right) = 4.914 \times 10^{-7} m$$

This radiation falls in the visible-light portion of the electromagnetic spectrum.

3-25 The Rutherford model of the atom does not specify where electrons might be found in the atom. The Bohr model does.

3-27 As the distance between the electron and nucleus increases the force holding them together decreases according to Coulomb's law.

3-29 (a) (ii) has the larger attractive force, $F = \dfrac{-4}{r^2}$

(b) (ii), $F = \dfrac{-1}{4r^2}$, is one fourth the force of (i) and (iii)

(c) According to Coulomb's Law $E = \dfrac{q_1 \times q_2}{r}$, so the one with the energy of smallest magnitude will be the easiest to separate. (i) and (ii) both have $E = \dfrac{-1}{r}$, while (iii) has a larger energy $E = \dfrac{-2}{r}$

3-31 $\lambda = \left(\dfrac{6.626 \times 10^{-34} J \ s \ \times \ 2.998 \times 10^8 m/s}{2.178 \times 10^{-18} J} \right) = 91.2$ nm. This is the wavelength of light required to move an electron from n=0 to n=∞. The electron will be completely removed from the atom.

3-33   From Fig. 3.5 the difference in energy between n=2 and n=3 is
        (-145.8 kJ/mol)-(-328.0 kJ/mol)=182.2 kJ/mol

$$182.2 \tfrac{kJ}{mol} \times 1000 \tfrac{J}{kJ} \times \frac{1 \ mole}{6.022 \times 10^{23} \ atoms} = 3.025 \times 10^{-19} \ J$$

$v=E/h$

$$v = \frac{3.025 \times 10^{-19} \ J}{6.626 \times 10^{-34} J \ s} = 4.565 \times 10^{14} \ \frac{cycle}{s}$$

3-35 $2.09 \times 10^{-18} \ J \ \times \ 6.022 \times 10^{23} \tfrac{atoms}{mol} \times \frac{1 kJ}{1000 J} = 1259 \tfrac{kJ}{mol}$ is the amount of energy absorbed by

the electron. If it starts at n=1 (E=-1312 kJ/mol), the final energy will be
(-1312 kJ/mol)-(-1259 kJ/mol)=-53 kJ/mol. According to Fig. 3.5 this is the energy of the
n=5 state. So the electron will have been promoted from n=1 to a final state of n=5.

3-37   First ionization energies increase from left to right across a row and decrease down a
        column in the periodic table. Na < Li < Be < F

3-39   Na:+1   Mg: +2   Al:+3   Si: +4   P: +5   S: +6   Cl: +7   Ar: +8
        K:+1   Ca:+2

3-41   Both Cl⁻ and Ar have the same number of electrons, however the Cl⁻ electrons only see a
        core charge of +7, whereas the electrons in Ar see a core charge of +8. Therefore the Ar
        will have a higher ionization energy than the Cl⁻.

3-43   The electron from the innermost shell is harder to remove because a) it is closer to the
        nucleus and b) it has no electrons shielding it, so the charge it sees is the full nuclear charge
        of +3, whereas the outer shell electron only sees a core charge of +1.

3-45   Because the outer-shell electrons in chlorine are farther from the nucleus than the outer
        shell electrons of fluorine.

3-47   The ionization energy of F⁻ would be less than that of Ne because F⁻ has one less proton in
        its nucleus than Ne.

3-49   Both Be and He have the same core charge on their two outer electrons. However, the
        outer electronic shell of Be is farther away from the nucleus than for He. Therefore, Be will
        have the lower ionization energy.

3-51   Generally the first ionization energies decrease going from top to bottom of a column of the
        periodic table.

3-53   The first ionization energy of hydrogen is larger than the first ionization energy of sodium
        because the $1s^1$ electron of hydrogen is closer to the nucleus and is therefore more tightly
        held by the nucleus. The core charge on the two atoms is the same.

3-55   Going down a column the core charge felt by the outer electrons is the same, however the
        size of the outer shell increases. Increasing the distance between the electron and nucleus
        reduces the coulombic attraction, lowering the ionization energy.

3-57   The element with the smallest first ionization energy is Ca.

3-59   In the first three rows, the core charge, number of valence electrons and group number are all the same.

3-61   No, the photoelectron spectrum of elements in the second row shows that electrons in the second shell will have two different energies after the first two electrons (for Li and Be) are placed in the shell. That is that the valence electrons for B through Ne reside in subshells with two slightly different energies.

3-63   No. The first ionization energy for Na is 495.8 kJ/mol. The radiation does not have the energy to ionize Na.

3-65   IE= hv-KE,   800.6 kJ/mol

3-67   Because both elements only have electrons in the 1s subshell.

3-69   The largest ionization energy in a PES is for the core electrons. The core electrons in B see a positive charge from the nucleus of +5, whereas the core electron in H, which is the only electron, sees only a charge of +1. Since the coulombic forces are greater with higher charge, the ionization energy will be higher for the core electrons of B than for H.

3-71   Electrons are found in three different subshells, 1s, 2s and 2p.

3-73   The two peaks at the right of the spectrum have similar energies but are quite different in energy from the first peak. This would suggest that these two peaks represent electrons from the same shell. The first peak represents the largest amount of energy needed to remove an electron and therefore corresponds to the electrons in the first shell that are most tightly held by the nucleus.

3-75   Mg. The energies are given in Table 3.6. The element has three shells. The first two are completely filled. The third shell has only the first sub-shell occupied. It has the same intensity as the other first sub-shells, therefore there are two electrons in it.

3-77   The element is Kr. The peaks shown correspond to the n=2, 3 and 4 shells. The electron configuration is $1s^2 2s^2 2p^6 3s^2\ 3p^6 3d^{10} 4s^2 4p^6$

3-79   (a) The element is P. There are 5 electrons in the valence shell and there are three shells.

(b)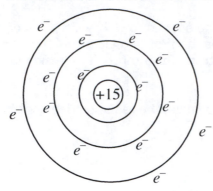

(c)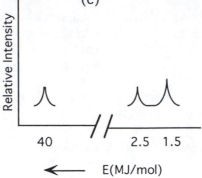

(d) $\lambda = 1.20 \times 10^{-8}$ m

$$v = 2.998 \times 10^8 \ \frac{m}{s} \times \left(\frac{1 \ \text{cycle}}{1.20 \times 10^{-8} \ m} \right) = 2.498 \times 10^{16} \ \frac{\text{cycle}}{s}$$

$E = h\nu$

$$E = (6.626 \times 10^{-34} \ J \ s) \left(2.498 \times 10^{16} \ \frac{\text{cycles}}{s} \right) \times \left(6.022 \times 10^{23} \ \frac{\text{atoms}}{\text{mol}} \right) \times \frac{MJ}{10^6 \ J} = 9.97 \ \text{MJ/mol}$$

the 3s and 3p electrons could be removed.

3-81   a) If l=1 then m_l can have values of -1,0, +1
        b) If n=2 then l can have values of 0 or +1

3-83   If n=4, then l = 0, 1, 2, 3. All orbitals of n=4 are in the same shell. The different l values represent both different angular shapes and subshells.

3-85   l= 0, s orbital; l =1, p orbital; l = 2, d orbital; l = 3, f orbital.

3-87   (c) and (e)

3-89   An unpaired electron will produce a small magnetic field. So if an atom has any unpaired electrons the atom may be magnetic.

3-91   (a) Atoms would be deflected in two separate areas.
        (b) All atoms would be deposited in the same area – no deflections in a magnetic field.
        (c) Paired electrons, no deflections
        (d) There are two unpaired electrons in F, so there are three possible net magnetic spins, +1, 0 and –1. This means the atoms will be deflected into 3 separate areas.

3-93   (b)

3-95   5, one for each value of m_l

3-97   Quantum numbers using selection rules outlined in the problem:
        n=1; l=0,1; m_l= 0, 1; m_s = +1, -1
        n=2; l=0, 1, 2; m_l=0, 1; m_s = +1, -1

3-99   When n=4 and l=3, the f set of orbitals is described. Since there are 7 values of m corresponding to l = 3 and two electrons can occupy each orbital, 14 electrons can have the quantum numbers n=4 and l=3.

3-101   Two. For any given n, l, and m_l values, m_s can only have the values +1/2 and -1/2.

3-103

n	max number of electrons
1	2
2	8
3	18
4	32

3-105   The last electron in Ga is found in a 4 p orbital. Answer (d).

3-107   To be in the same orbital two electrons must have the same values of n, l, and m_l, but they must have opposite values of spin, m_s.

3-109   (e) is incorrect.

3-111   On the basis of electron configuration hydrogen ($1s^1$) should be in Group 1 (IA) along with lithium ($[He]2s^1$). Since hydrogen is the only non-metal in that group, it would also be appropriate to include it in group VIIIA with the non-metals that have one less electron than a filled-shell configuration. Helium with a $1s^2$ electron configuration should be placed in Group 2 (IIA). Since helium is a non-metal and behaves more like the elements with filled-shell electron configuration it is placed in Group 18 (VIIIA).

3-113   Five electrons beyond the element number 36, Kr is the element number 41, niobium, 41, found in row 5 and column 5.

3-115   Element 119 should belong to Group IA of the periodic table if and when it is discovered.

3-117   N $[He]\ 2s^2\ 2p^3$

3-119   x=2

3-121   (c) is correct for the electronic configuration of the Br$^-$ ion.

3-123   6 electrons in s orbitals in the Ti^{2+} ion

3-125   (d) Y is the first element to have 4d electrons in its electronic configuration.

3-127   (b) satisfies Hund's rules for the electronic configuration of carbon.

3-129   (a), (b) and (d) are incorrect. Only (c) is correct.

3-131   (d) Fe^{3+} has five unpaired electrons.

3-133   The size of an atom increases as we go down a column. As we go down a column of the periodic table, electrons are placed in shells farther from the nucleus. When this happens, the size of the atom increases.

3-135
$$1.442 \ \overset{\circ}{A} \times \frac{10^{-8} \ cm}{1 \ \overset{\circ}{A}} \times \frac{1 \ m}{100 \ cm} \times \frac{10^{9} \ nm}{1 \ m} = 0.1442 \ nm$$

$$1.442 \ \overset{\circ}{A} \times \frac{10^{-8} \ cm}{1 \ \overset{\circ}{A}} \times \frac{1 \ m}{100 \ cm} \times \frac{10^{12} \ pm}{1 \ m} = 144.2 \ pm$$

3-137 The covalent radius, the distance between adjacent atoms in a covalent bond is smaller than the metallic radius because the sharing of electrons in the covalent bond tends to attract atoms more closely together. The atoms of a metal are found in an array of extended planes of atoms and the force of attraction between the nuclei of these adjacent atoms is less, hence the metallic radius is larger than the covalent radius.

3-139 The covalent radius decreases when going from left to right in a single row since the electrons are being placed in the same shell, but the core charge experienced by the electrons increases. In going down a single column on the periodic table, the covalent radius increases since the size of the outer shell is increasing.

3-141

	Mg	S
Covalent radius (nm)	0.136	0.104
Ionic radius (nm)	0.065 (Mg^{2+})	0.184 (S^{2-})

Magnesium ions are smaller, because they have given up their $3s^2$ electrons in forming Mg^{2+} ions. The electrons remaining are in the smaller 1s and 2s, 2p orbitals. The sulfur ion S^{2-} has increased its electrons by 2 while remaining in the same 3p orbitals. This results in S^{2-} having a larger ionic radius than Mg^{2+}.

3-143 The ion with the fewer electrons will be smaller. $Fe^{3+} < Fe^{2+}$

3-145 The size of negative ions increases down a column. The order of increasing ionic radius is predicted to be: $H^- < F^- < Cl^- < Br^- < I^-$
The ionic radii (in nm) as given in the Appendix are:

0.136 <	0.208 <	0.181 <	0.196 <	0.216
F^- <	H^- <	Cl^- <	Br^- <	I^-

The exception is hydrogen where the ratio of electrons to protons is much higher than those of the other elements in this series.

3-147 (e) Se^{2-} because it's the only one in the fourth row

3-149 (c) Be^{2+} because it has the largest positive charge of the second row elements

3-151 (b) P^{3-} It has the smallest number of protons holding the electrons.

3-153 More highly charged cations have higher ionization energies. P^{4+} of the species listed will have the largest ionization energy.

3-155 The great increase in IE between the third and fourth ionization energies suggests an element in period three with 3 electrons in that period. This would be aluminum with electron configuration [Ne] $3s^2 \ 3p^1$, response d.

3-157 (a) Of the elements presented, Na has the largest second ionization energy.

3-159   The order of increasing second ionization energy is
         Mg < Be < Ne < Na < Li

3-161   The AVEE of Be is 0.90 as compared to 1.89 for oxygen. Since the AVEE is a measure of
         how tightly an atom holds its valence electrons, it is more difficult to remove electrons
         from O as compared to Be.

3-163   (a) Mg < P < Cl
         (b) Se < S < O < F
         (c) K < P < O

3-165   From left to right on the periodic table AVEE will increase. This is because the first (and
         subsequent ionization energies) increase in going from left to right.

3-167   As the value n becomes larger the energy difference between subshells becomes smaller.

3-169   Nonmetals have large values for AVEE.

3-171   Carbon has a high AVEE (greater than 1.26mJ/mol), making it a nonmetal. The AVEE of Si
         is 1.13 MJ/mol putting in region of a semimetal. Sn has a low AVEE making metallic.

3-173   Tl < Ga < Al < B

3-175   As an atom gets larger, the outer electrons are easier to remove because the distance of
         the electrons from the nucleus increases (see Coulomb's Law).

3-177   Increasing Ionization Energy: O^{2-} < F^- < Ne < Na^+ < Mg^{2+}
         Increasing Radius: Mg^{2+} < Na^+ < Ne < F^- < O^{2-}
         Increasing Ionization Energy: Na < Mg < O < F < Ne
         Increasing Radius: Ne < F < O < Mg < Na

3-179   Relative ionization energies for an isoelectronic series depend on the total number of
         protons present. The greater the number of protons, the more strongly the outermost
         electron is held, and the higher the first ionization energy. In this case, the order of
         increasing ionization energy is S^{2-}(16 protons) < Ar(18 protons) < K^+(19 protons). In an
         isoelectronic series, the cations are smaller than the neutral atoms which are in turn
         smaller than the anions. Thus, the order of increasing radii would be K^+ < Ar < S^{2-}.

3-181   (a) C, 2 unpaired electrons (b) N, 3 unpaired electrons (c) O, 2 unpaired electrons
         (d) Ne, 0 unpaired electrons (e) F, 1 unpaired electron

3-183   In the Stern-Gerlach experiment beams of atoms will interact with a magnetic field and
         split into two separate beams if they possess unpaired electrons. The magnetic moment
         is related to the number of unpaired electrons. Since He and Ne do not have a magnetic
         moment, we would expect that beams of these atoms would not be split. Beams of the
         other atoms should be split in two because they have magnetic moments.

3-185   Element Z is diamagnetic and typically forms +2 cations. This is consistent with Z having
         2 electrons in its outermost shell. Group 2A matches this description. Since Z has the
         next to lowest ionization energy in its group and ionization decreases down a group, Z
         must be the next to last element in Group 2A: Ba. The compounds formed would be BaO
         and $BaCl_2$.

3-187

Atom	Chemical Symbol	Magnetic field behavior	First IE	Atomic Radius	Number of PES Peaks	Core Charge	AVEE	Number of Valence Electrons
X	Ne	not deflected	2.08	0.070	3	+8	2.0	8
Y	Na	deflected	0.50	0.16	4	+1	0.50	1
Z	Mg	not deflected	0.74	0.14	4	+2	0.8	2

3-189   (a) Cl. The atom is in the group VIIA with a core charge of +7 and very high AVEE indicates it is high up in that column. And the covalent radius matches other atoms in the third row.

(b) Cl: $1s^22s^22p^63s^23p^5$

(c) Cl$^-$: $1s^22s^22p^63s^23p^6$. The radius of Cl$^-$ would be larger since there is one more electron, but the same core charge.

(d) It is easier to ionize Cl$^-$ than it is Cl, because the number of electrons is larger and the core charge is the same.

(e) AVEE (Cl) < AVEE (Ar). AVEE increases to the right of the periodic table.

(f) This element is F. With a higher AVEE, it means the electrons are more tightly bound, probably in a smaller shell, indicating a smaller covalent radius.

(g) 7

3-191   (a) (ii) matches the ionization energies; two low energies followed by two higher energies from the next shell.

(b) 2   the two in the outer shell.

(c) +2

(d) You are taking an electron from something which is already a positive ion.

Chapter 4
The Covalent Bond

Note to students:

There are several schools of thought on what constitutes the best Lewis structures for molecules. The best evidence, of course, is experimental data. In the absence of data, some chemists prefer to minimize the formal charge in the structure. Thus, for example, for SO_2, the structure can be drawn with one or two double bonds. In one case, sulfur has 10 valence electrons and no formal charge; in the other, sulfur has eight valence electrons and a formal charge of +1. Without additional information, it is not possible to feel confident of the correct structure. A recent study[1] of Lewis structures suggests that multiple bonds (expanded octets) may not be the best representation of the structures of many molecules commonly drawn this way. We have generally chosen not to use expanded octets in writing the structures of this text.

4-1 Valence electrons are the electrons in an atom's outermost shell. This means they are the electrons on an atom that were not present in the previous Group VIIIA (Group 18) element, ignoring filled d or f subshells.

4-3 (a) Fe has 8 valence electrons.
(b) Zr has 4 valence electrons.
(c) Bi has 5 valence electrons.
(d) I has 7 valence electrons.

4-5 Na^+, Mg^{2+}, Al^{3+}, and Sc^{3+} all have 8 valence electrons. All of these elements lose electrons to achieve an octet of electrons.

4-7 The **octet rule** refers to Lewis's discovery that main-group elements will gain or lose electrons until they have eight electrons in their outermost shell.

4-9 For main-group elements and alkaline metals the group number is the same as the number of valence electrons.

4-11 In the covalent bond of F_2 there is a single pair of electrons shared between the two fluorine atoms. In the O_2 molecule there are two pairs of electrons shared between the two atoms. However, in both molecules there are eight valence electrons that surround each atom.

4-13 When two atoms are brought together the electrons on atom A would be attracted to the nucleus of atom B and *vice versa*. There would also exist an equal and opposite repulsion between the two nuclei and the separate electrons. These two forces would appear to negate each other as shown in Figure 4.4. However, if the shared electrons between the two atoms were specifically placed in the region between the two atoms the repulsions due to the nuclei can be minimized.

4-15 A **bonding domain** is a region of space that contains two spin paired electrons that are being shared between two atoms. The domain extends over both atoms, but effectively concentrates the electron density in the space between the atoms.

[1] L. Suidan, J.K. Badenhoop, E.D. Glendenins, and F. Weinhold, J. Chem. Ed. **72** 583(1995).

4-17   According to the model for the electron configurations of an atom, two and only two electrons can share a common region of space. The two electron occupants have mutual access to the attraction of the two bound nuclei (and find themselves in an environment much more favorable than that existing in the isolated atoms). Thus, the unpaired valence electrons of two chlorine atoms come together to form the covalent bond of Cl_2.

4-19

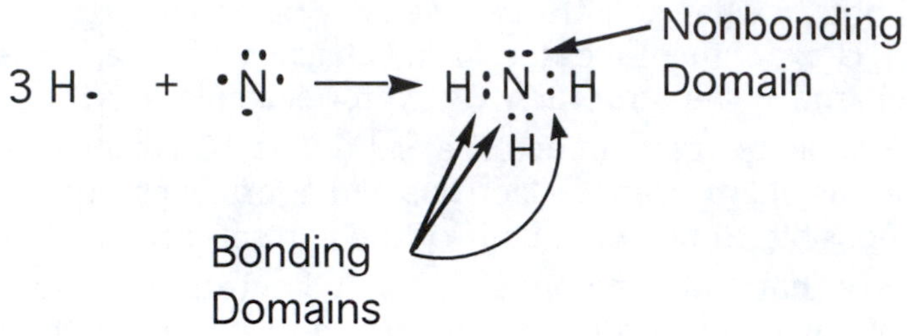

$$3\ H\cdot\ +\ \cdot\ddot{\underset{\cdot}{N}}\cdot\ \longrightarrow\ H\!:\!\ddot{N}\!:\!H$$

Nonbonding Domain

Bonding Domains

4-21   Each atom should have eight electrons surrounding it to satisfy the octet rule.

4-23   (c)

4-25   In general the element with the lowest AVEE will be the central atom. Since S is directly below O on the periodic table we automatically know that it has a lower AVEE than oxygen. Therefore structure (a) is the best.

4-27   (a)

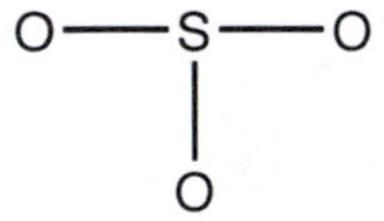

$$\begin{array}{c} O-\!\!-S-\!\!-O \\ | \\ O \end{array}$$

(b) O—S—O

(c) O—O—O

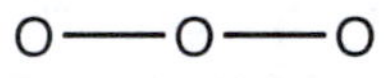

(d)

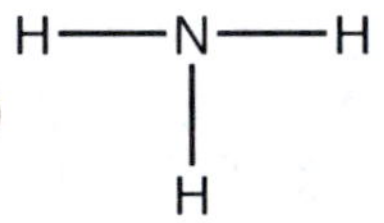

$$\begin{array}{c} H-\!\!-N-\!\!-H \\ | \\ H \end{array}$$

(e)

$$\begin{array}{c} H \\ | \\ Cl-\!\!-C-\!\!-Cl \\ | \\ Cl \end{array}$$

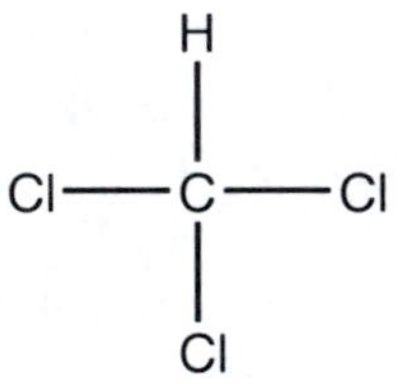

4-29   (a)   8 for Kr + 2(7) for F = 22 valence electrons
       (b)   6 for S + 4 (7) for F = 34 valence electrons
       (c)   4 for Si + 6(7) for F + 2 for charge = 48 valence electrons
       (d)   6 for S + 4(6) for O + 2 for charge = 32 valence electrons

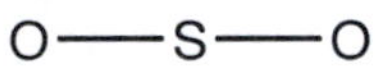

4-31

(a)

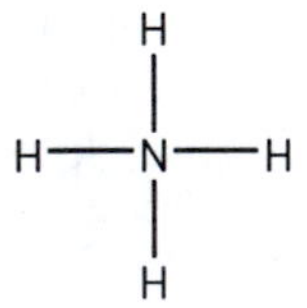

$$\begin{array}{c} H \\ | \\ H-\!\!-N-\!\!-H \\ | \\ H \end{array}$$

(b)

$$\begin{array}{c} O-\!\!-N-\!\!-O \\ | \\ O \end{array}$$

(c)

$$\begin{array}{c} O \\ | \\ O-\!\!-S-\!\!-O \\ | \\ O \end{array}$$

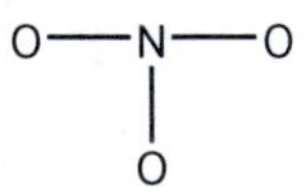

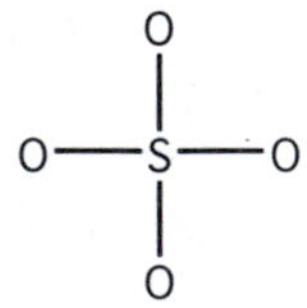

4-33 (a) NO_3^-

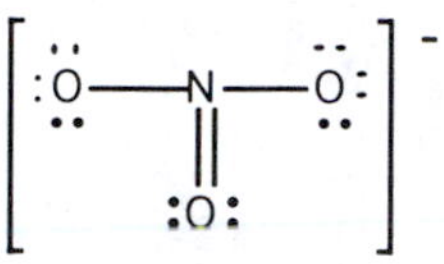

(b) SO_3^{2-}

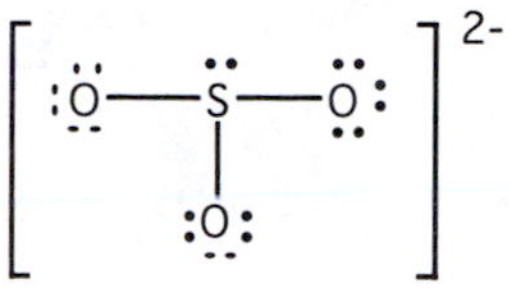

or

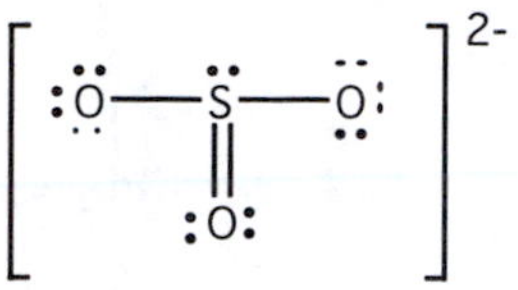

The structure with the minimum number of formal charges is generally preferred by chemists.

(c) CO_3^{2-}

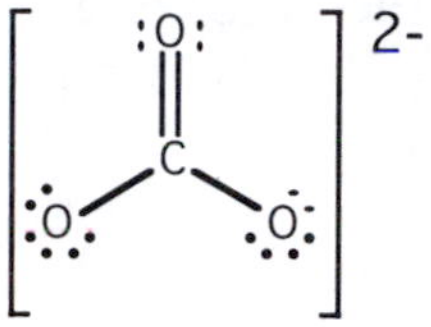

(d) NO_2^+

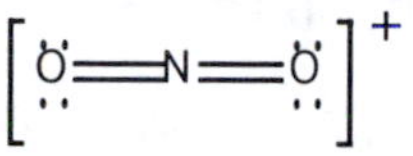

4-35 (a) N_2O

I   II

Structure I is preferred.

(b) N_2O_3

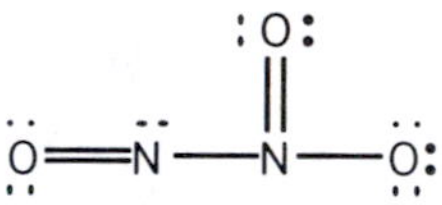

4-37 In the molecule N_2O_5, there are 40 valence electrons available. A structure that contains $O_2N\text{-}NO_3$ would require at least 42 valence electrons. The correct Lewis structure is:

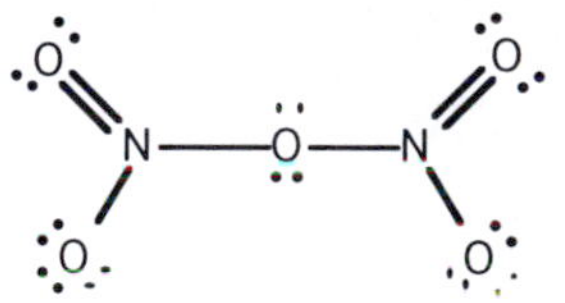

4-39 (a) SO_2 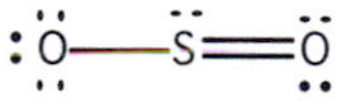   (b) SO_3 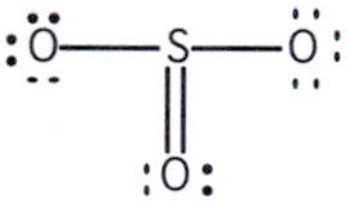   (c) SO_3^{2-} 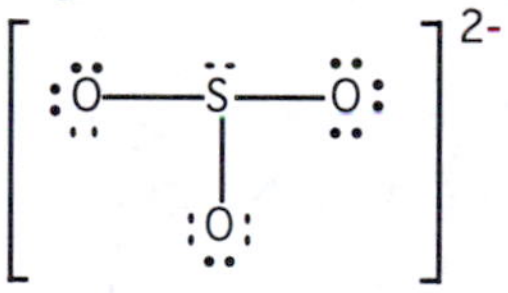   (d) SO_4^{2-}

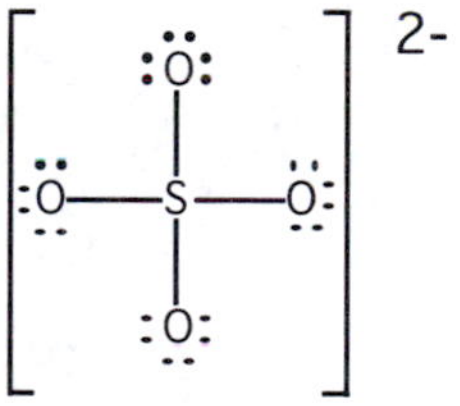

Structures (a),(b), and (d) have been shown to be of more significance than the expanded octet structures. See reference 1 at the beginning of this problem set.

4-41 The N_2 molecule has 10 valence electrons.
   (a)   CO has 10 valence electrons.
   (b)   NO has 11 valence electrons.
   (c) CN^- has 10 valence electrons.
   (d) NO^+ has 10 valence electrons.
   (e) NO^- has 12 valence electrons.
CO, CN^-, and NO^+ have the same electronic configuration as the N_2 molecule.

4-43   (a)   BF$_3$

Boron is surrounded by six valence electrons. This is an exception to the octet rule.

(b)   H$_2$CO

Formaldehyde is not an exception to the octet rule.

(c)   XeF$_4$

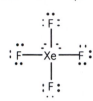

Xenon is surrounded by twelve valence electrons. This is an exception to the octet rule.

(d)   IF$_3$

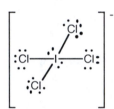

Iodine is surrounded by 10 valence electrons. This is an exception to the octet rule.

4-45   a)   b)   c)   d)

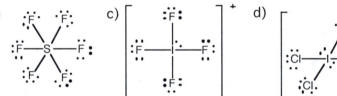

None of these structures obey the octet rule.

4-47   The shorter the bond length the higher the bond order; therefore N$_2$ is a triple bond, HNNH is a double bond and H$_2$NNH$_2$ is a single bond.

4-49   Since the bond lengths are about the same, the bond orders and hence the bond strengths must also be similar.

4-51   The resonance structures of SCN$^-$ are:

$$\left[\ddot{S}=C=\ddot{N} \right]^- \longleftrightarrow \left[:S\equiv C-\ddot{N}: \right]^- \longleftrightarrow \left[:\ddot{S}-C\equiv N: \right]^-$$

4-53   Responses (b) and (c) have no resonance structures.

4-55   Since AVEE increases from left to right across in the periodic table, electronegativity also increases. Moving from left to right along a row in the periodic table metallic character decreases and covalent character increases. Since covalent radii are smaller than metallic radii, it is harder to remove an electron from an atom that has a greater covalent character i.e., the electron is closer to the nucleus and more tightly held. This is also consistent with the AVEE data rationalizing the trend of increasing electronegativity.

4-57   Response (b) is correct.

4-59 (a)

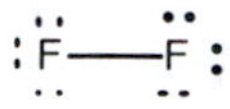

:F——F: $\delta_F = V_F - N_F - B_F\left(\dfrac{EN_F}{EN_F + EN_F}\right)$ $\delta_F = 7 - 6 - 2\left(\dfrac{4.19}{4.19 + 4.19}\right) = 7 - 6 - 1 = 0$

(b) 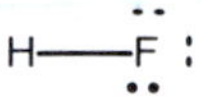 H——F: $\delta_F = 7 - 6 - 2\left(\dfrac{4.19}{4.19 + 2.30}\right) = 7 - 6 - 1.29 = -0.29$

(c) 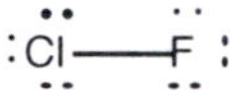 :Cl——F: $\delta_F = 7 - 6 - 2\left(\dfrac{4.19}{4.19 + 2.87}\right) = 7 - 6 - 1.19 = -0.19$

4-61   :B——F:

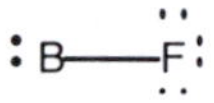

$\delta_F = 7 - 6 - 2\left(\dfrac{4.19}{4.19 + 2.05}\right) = -0.34$

$\delta_B = 7 - 6 - 2\left(\dfrac{2.05}{4.19 + 2.05}\right) = +0.34$

4-63

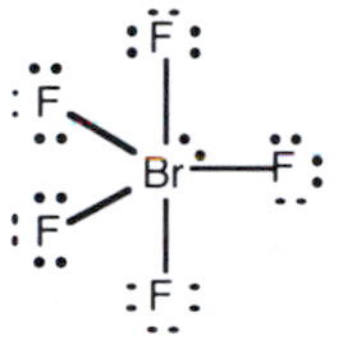

$\delta_O = 6 - 4 - 4\left(\dfrac{3.61}{3.61 + 2.54}\right) = -0.35$

$\delta_H = 1 - 0 - 2\left(\dfrac{2.30}{2.30 + 2.54}\right) = +0.05$

4-65   (a) H, lower electronegativity than C.
(b) C, lower electronegativity than Cl.
(c) B, lower electronegativity than H.
(d) N, lower electronegativity than O.
(e) N, lower electronegativity than both Cl and O and it's between the two.

4-67   The formal charge on an atom is the difference between its number of valence electrons and the number of electrons the atom formally has in the Lewis structure. This is calculated by dividing the pairs of electrons in each covalent bond between the atoms (one to each atom) and then comparing the number of electrons that are now formally assigned to each atom with the number of valence electrons on the neutral atom.
(a) HBr

Formal   H——Br: charge (Br) = 7 valence e⁻ in neutral atom - 7 e⁻
                            formally assigned = 0

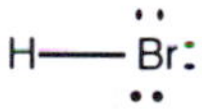

(b) Br₂   :Br——Br:

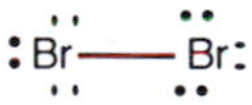

7   Formal charge (Br) = 7 valence e⁻ in neutral atom - 7

                    formally assigned = 0

(c)   HOBr

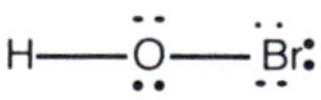

H——O——Br: Formal charge (Br) = 7 valence e⁻ in neutral atom -7 e⁻
                formally assigned = 0
                (d)   BrF₅

                        Formal charge (Br) = 7 valence e⁻ in neutral atom -7 e⁻
                        formally assigned = 0

4-69 The formal charge on an atom is the difference between its number of valence electrons and the number of electrons the atom formally has in the Lewis structure. This is calculated by dividing the pair of electrons in each covalent bond between the atoms (one to each atom) and then comparing the number of electrons that are now formally assigned to each atom with the number of valence electrons on the neutral atom.

(a) N_2O — For the most important resonance structure

Formal charge on the end (N) = 5 valence e⁻ in neutral atom -5 e⁻ formally assigned = 0

Formal charge on middle (N) = 5 valence e⁻ in neutral atom -4 e⁻ formally assigned = +1

(b) N_2O_3

Formal charge (N_1) = 5 valence e⁻ in neutral atom - 5 e⁻ formally assigned = 0

Formal charge (N_2) = 5 valence e⁻ in neutral atom -4 e⁻ formally assigned = +1

(c) N_2O_5

4 e⁻ Formal charge (N) = 5 valence e⁻ in neutral atom -4 e⁻ formally assigned = +1

4-71 The formal charge on an atom is the difference between its number of valence electrons and the number of electrons the atom formally has in the Lewis structure. This is calculated by dividing the pair electrons in each covalent bond between the atoms (one to each atom) and then comparing the number of electrons that are now formally assigned to each atom with the number of valence electrons on the neutral atom.

$S_2O_3{}^{2-}$

Formal charge end (S) = 6 valence e⁻ in neutral atom -7 e⁻ formally assigned = -1

Formal charge middle (S) = 6 valence e⁻ in neutral atom -4 e⁻ formally assigned = +2

Formal charge each (O) = 6 valence e⁻ in neutral atom -7 e⁻ formally assigned = -1

or

Formal charge end (S) = 6 valence e⁻ in neutral atom - 6e⁻ formally assigned = 0

Formal charge middle (S) = 6 valence e⁻ in neutral atom - 6 e⁻ formally assigned = 0

Formal charge (single bond O) = 6 valence e⁻ in neutral atom -7 e⁻ formally assigned = -1

Formal charge (double bond O) = 6 valence e⁻ in neutral atom - 6 e⁻ formally assigned = 0

4-73

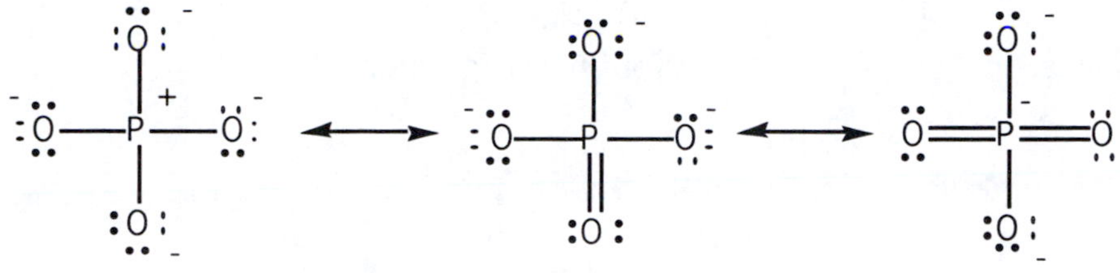

Structure II would be the preferred structure. The formal charge on each atom is minimized, and the negative formal charge is on the more electronegative oxygen. Experimental measurements of bond lengths could be used to confirm this. See however reference 1.

4-75

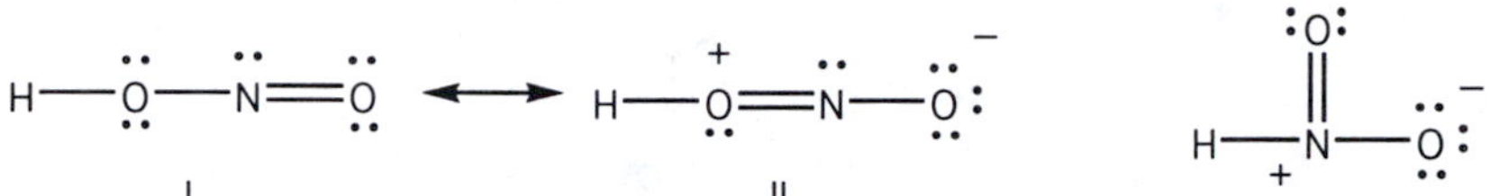

The first structure, HONO, has two possible Lewis structures. The one on the left, I, has no formal charge on any atom. There is only one structure for HNO_2. This structure has a formal charge on N and one O. Structure I is best.

4-77   The three structures are

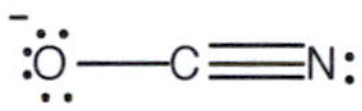

Oxygen is the most electronegative atom in this molecule so a negative charge should be placed at that atom, making structure c) the best structure.

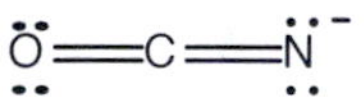

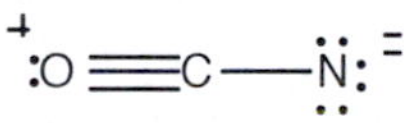

4-79

a) 3 domains, trigonal planar, 120°.

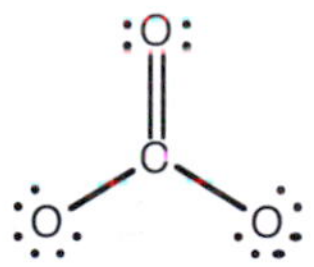

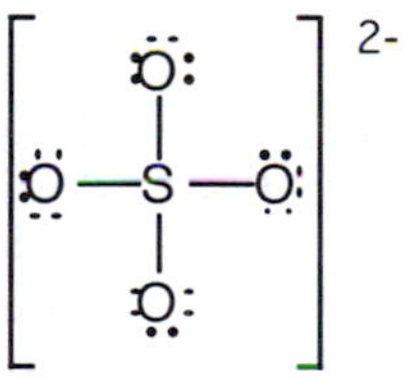

b)   4 domains, tetrahedral, 109.5°

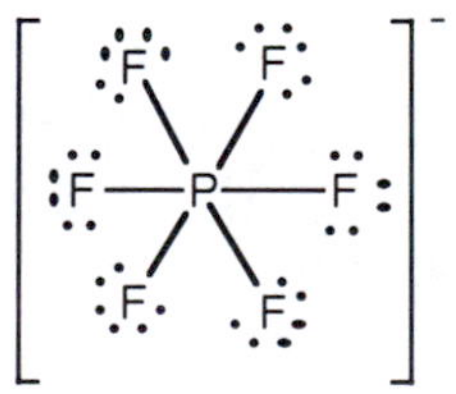

c)   6 domains, octahedral, 90°

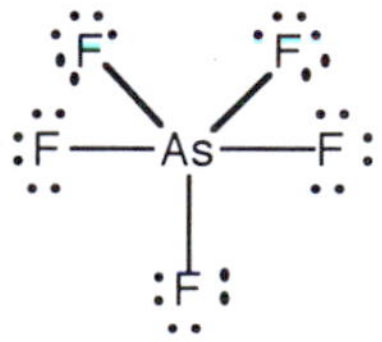

d) 5 domains, trigonal bipyramid, 90° and 120°
120°

33

4-81　(a)　NH$_4^+$
four bonding domains,

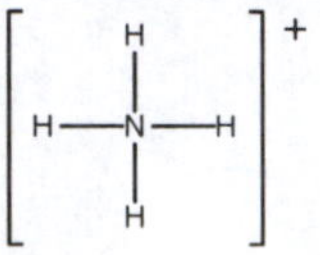

$$\left[\begin{array}{c} H \\ | \\ H-N-H \\ | \\ H \end{array} \right]^+$$

bond angles of 109.5°

(b)　HCCl$_3$
four bonding domains,

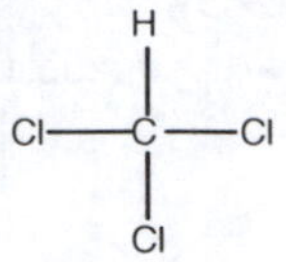

$$\begin{array}{c} H \\ | \\ Cl-C-Cl \\ | \\ Cl \end{array}$$

bond angles of 109.5°

(c)　BeH$_2$
two bonding domains,

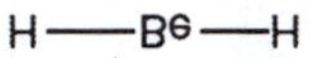

H——Be——H

bond angles of 180°

(d)　OCCl$_2$
three bonding domains,

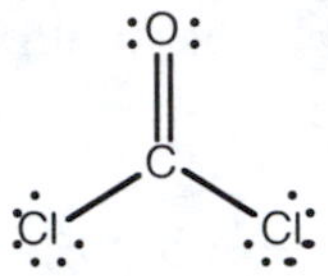

bond angles of 120°

(e)　OCS

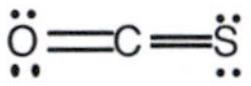

Ö=C=S̈

two bonding domains,
bond angles of 180°

4-83　(a) OH$^-$

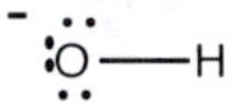

Three pairs of nonbonding electrons

(b)　O$_2$

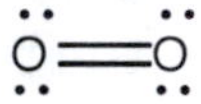

Ö=Ö

Four pairs of nonbonding electrons

(c) CO$_3^{-2}$

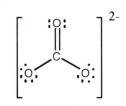

Eight pairs of nonbonding electrons

(d)　Br$^-$

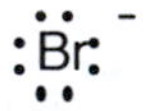

:Br: $^-$

Four pairs of nonbonding electrons

(e) NH$_3$

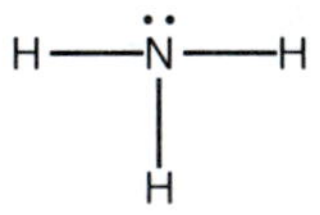

H——N——H
|
H

One pair of nonbonding electrons

4-85

Bonding domains	Nonbonding domains	Arrangement	Molecular geometry
2	0	linear	linear
4	0	tetrahedral	tetrahedral
2	2	tetrahedral	bent
5	0	trigonal bipyramidal	trigonal bipyramidal
3	2	trigonal bipyramidal	T-Shaped
5	1	octahedral	square pyramid
4	2	octahedral	square planar

4-87 (a) PO_4^{3-}

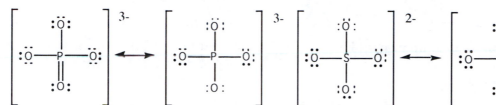

Number of bonding domains= 4
Number of nonbonding domains = 0
Geometry = tetrahedral

(b) SO_4^{2-}

Number of bonding domains= 4
Number of nonbonding domains = 0
Geometry = tetrahedral

(c) XeO_4

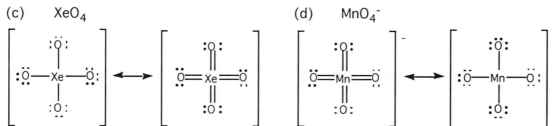

Number of bonding domains = 4
Number of nonbonding domains = 0
Geometry = tetrahedral

(d) MnO_4^-

Number of bonding domains = 4
Number of nonbonding domains = 0
Geometry = tetrahedral

4-89 (a) SF_3^+

Number of bonding domains= 3
Number of nonbonding domains = 1
Geometry = trigonal pyramidal

(b) SF_4

Number of bonding domains = 4
Number of nonbonding domains = 1
Geometry = see-saw

(c) SF_5^-

Number of bonding domains= 5
Number of nonbonding domains = 1
Geometry = square pyramidal

(d) SF_6

Number of bonding domains= 6
Number of nonbonding domains = 0
Geometry = octahedral

4-91 (a) N$_2$O

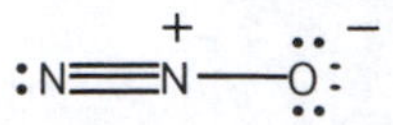

:N≡≡N⁺—Ö⁻:

Number of bonding domains = 2
Number of nonbonding domains= 0
Geometry = linear

(b) NO$_2^-$

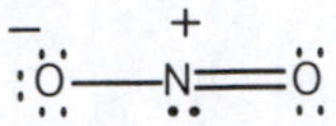

:Ö⁻—N⁺==Ö

Number of bonding domains = 2
Number of nonbonding domains= 1
Geometry = bent

(c) NO$_3^-$

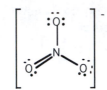

Number of bonding domains = 3
Number of nonbonding domains = 0
Geometry = trigonal planar

4-93 Response (a). XeF$_3^+$ has a T-shaped molecular geometry.

CH$_3$—Hg—CH$_3$

4-95 SF$_4$ seesaw   N$_2$O   linear
      CH$_4$ tetrahedral   NO$_2$ bent
      CO$_2$ linear   PCl$_4^+$ tetrahedral
      H$_2$O   bent   PCl$_4^-$ seesaw
      BeH$_2$ linear
      Response (c) CO$_2$ and BeH$_2$ are linear.

4-97 (a)   SO$_3$ trigonal planar   (d)   PF$_3$ trigonal pyramidal
      (b)   SO$_3^{2-}$ trigonal pyramidal   (e)   BH$_3$ trigonal planar
      (c)   NO$_3^-$ trigonal planar
      Responses (a),(c),and (e) have planar geometry.

4-99 (a)   C$_2$H$_2$ linear   (d)   NO$_2^+$ linear
      (b)   CO$_2$ linear   (e)   H$_2$O   bent
      (c)   NO$_2^-$ bent
      Responses (a), (b), and (d) have linear geometry.

4-101

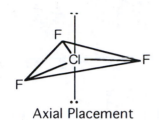

Axial Placement 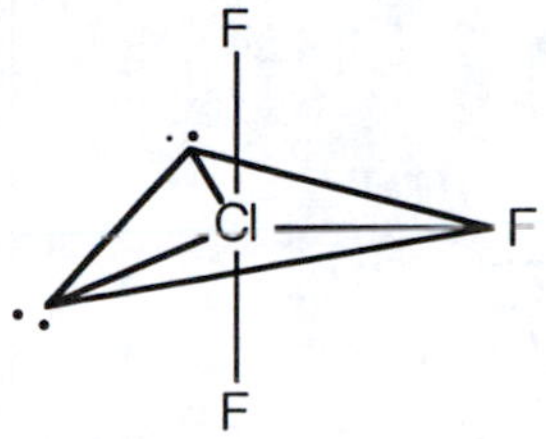   Equatorial Placement

With the nonbonding domains in the equatorial position, there are four unfavorable 90°
bonding-nonbonding interactions. With the nonbonding domains in the axial positions,
there are 6 unfavorable 90° bonding-nonbonding interactions. The placement of the
nonbonding domains in the equatorial position gives a more stable structure.

4-103

Bonding Domains	Nonbonding Domains	Bond Angle
2	0	180°
5	0	90° and 120°
3	0	120°
2	1	<120°
2	3	180°
4	0	109.5°
3	1	<109.5°
2	2	<109.5°
2	3	180°
3	2	<90°
4	2	90°

4-105   a) both C 109.5°,   O <109.5°
          b) 109.5°
          c) O <109.5°,   N <120°
          d) H_3C < 109.5°,   O=C 120°
          e) 120°

4-107 a)

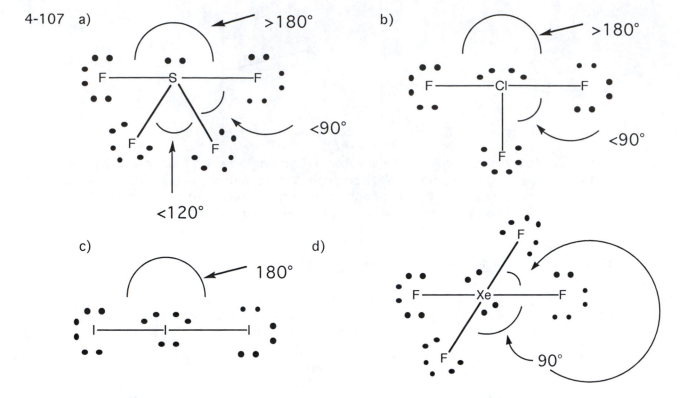

4-109 The carbon-sulfur bond is a polar bond. In CS_s the dipoles cancel because the molecule is symmetrical. Therefore CS_2 is not a polar molecule. The dipoles created by the bonds in OCS, however, do not cancel.

4-111 Responses (a),(b), (c), and (d) are polar.

4-113 The Lewis structures of thionyl chloride ($SOCl_2$) and sulfuryl chloride (SO_2Cl_2) are:

thionyl chloride   sulfuryl chloride

Thionyl chloride has a dipole moment along the S-O bond. Therefore, it is a polar molecule. There are two possible structures that could be drawn for sulfuryl chloride and both structures have a tetrahedral arrangement. Since there are two types of polar bonds in the molecule, S-Cl and S-O, the individual bond dipoles do not cancel. Sulfuryl chloride is a polar molecule.

4-115

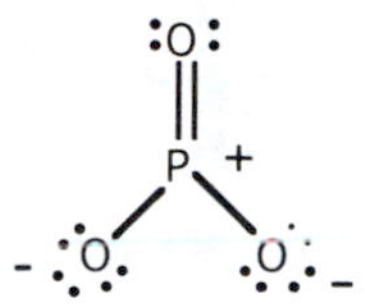

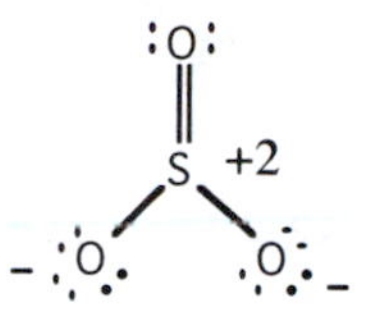

Phosphorous with a formal charge of +1 is capable of interactions with an oxygen, formal charge -1, of another PO_3^- unit to polymerize. Similarly SO_3 would be expected to polymerize.

4-117 (a)   N O   (b)   N O_2 (c) ClO_2 (d) ClO_3

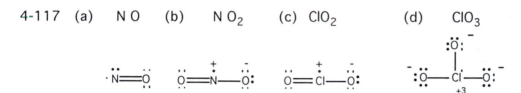

All of these molecules have an unpaired electron in the Lewis structure.

4-119 In the molecule with the Lewis structure,

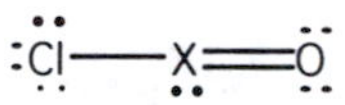

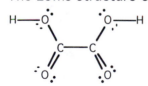
, the central atom must have five valence electrons due to formal charge considerations. The element is N. Response (d).

4-121 For the molecule XF_4^-, F will not double bond to the central atom. The central atom must have 4 valence electrons for bonding to the four fluorine atoms. In addition, the central atom needs another 4 valence electrons to account for the 2 pairs of nonbonding electrons. The molecule has a -1 charge, so we must subtract 1. The number of valence electrons on the central atom is 4 + 4 - 1 = 7. The central atom belongs to Group VIIA.

4-123 Response (a). N_2 has a triple bond. All the others have a double bond.

4-125 The Lewis structure of oxalic acid is:

4-127 There are many possible Lewis structures for this molecule. Only those that maintain octets on all the atoms will be shown.
(a)

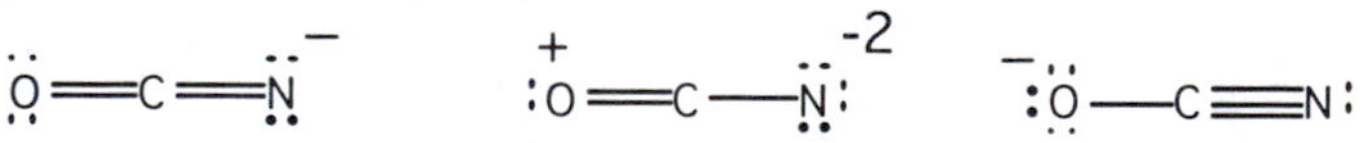

(b) III minimizes formal charges and places -1 formal charge on the most electronegative atom.
(c) If III is the major contributor, followed by I, then on average the C-O bond length should be between 133 and 150 pm. Dipole moment data could be used. If structure III is the major contributor, then the dipole moment should point towards the O along the C-O bond axis.
(d) Since C has 2 bonding domains and 0 nonbonding pairs, the geometry is linear.
(e) Yes; the bond dipoles do not cancel.

4-129 Thymine

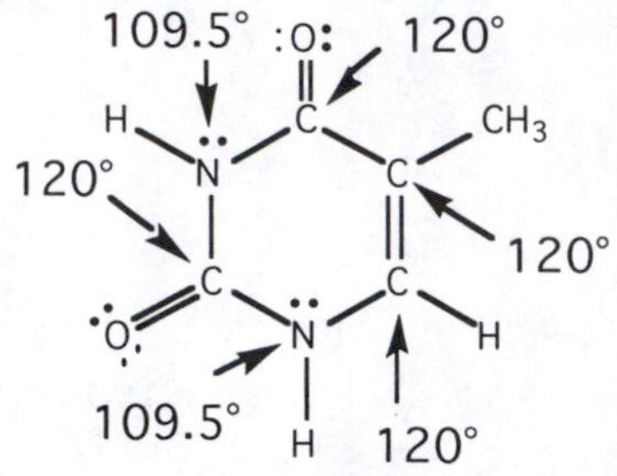

4-131 (a) Nitrogen has an incomplete octet.
(b) Too many electrons are shown.
(c) All valence electrons are not shown.
(d) Hydrogen should only share two electrons.
(e) Hydrogen should only share two electrons.

Chapter 4 Special Topics

4A-1 Orbital overlap occurs when two orbitals, one from each atom engaged in a bond, occupy the same place in space.

4A-3 NH_3

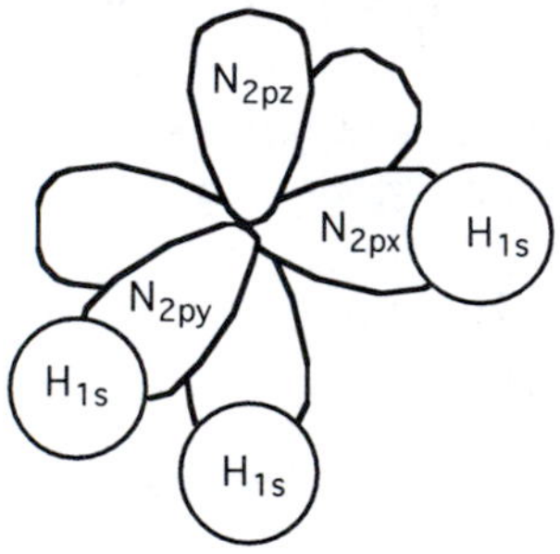

H_2S

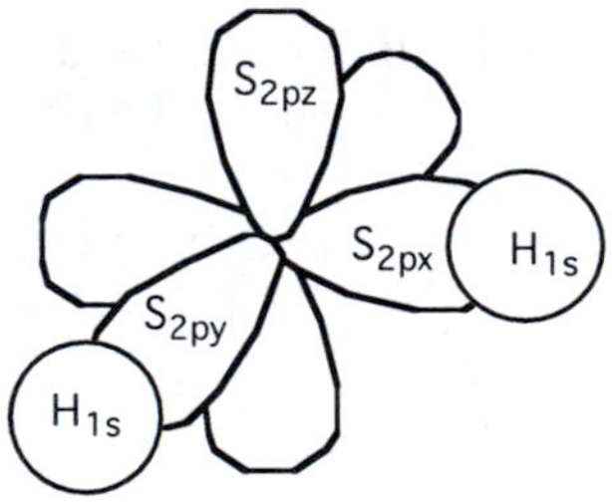

NO_2

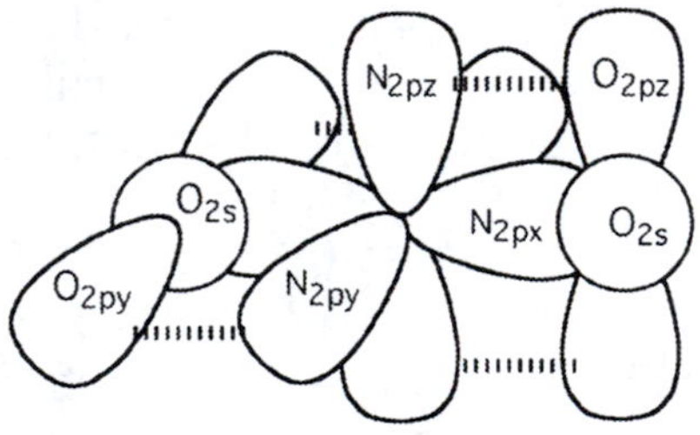

C_2

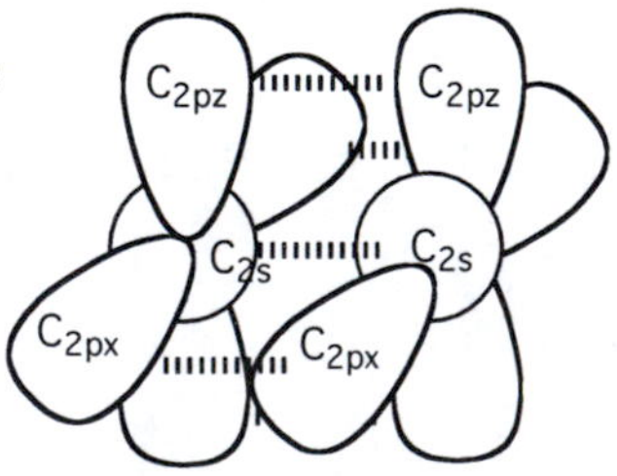

4A-5 (a) CH_4 sp^3
(b) H_2CO sp^2
(c) HCO_2^- sp^2

4A-7 The Lewis structure for the molecule ethylene is

$$\begin{array}{cc} H & H \\ \diagdown & \diagup \\ C = C \\ \diagup & \diagdown \\ H & H \end{array}$$

The hybridization on each carbon is sp². Each C-H bond is σ (sp²-s). One of the C-C bonds is a σ (sp²-sp²) and the other is a π (2p-2p). The molecule is planar.

The Lewis structure for the molecule acetylene is:

$$H - C \equiv C - H$$

The hybridization on each carbon is sp. The C-H bonds are σ (sp-s) bonds. One C-C bond is σ (sp-sp) and the other two C-C bonds are π (2p-2p) bonds. The molecule is linear.

4A-9 A σ molecular orbital is one that results from the head-on overlap of atomic orbitals, that is the electron density lies along the internuclear axis. A π molecular orbital is one that results from the sideways or parallel overlap of atomic orbitals. The electron density in a π molecular orbital does not lie on the internuclear axis.

4A-11 The stronger the interaction between a pair of atomic orbitals, the larger the difference between the energies of the bonding and antibonding orbitals formed. The σ_p molecular orbitals result from the head-on overlap of $2p_z$ atomic orbitals. This is a stronger interaction than the sideways overlap of the $2p_x$ and $2p_y$ atomic orbitals that results in the π_x and π_y molecular orbitals. Therefore the difference in energy between the σ_p and the σ^*_p will be greater than the difference in energy between the π_x π_y and π^*_x π^*_y molecular orbitals.

4A-13 The bond order (BO) for a diatomic molecule is calculated by placing the valence electrons in the molecular orbitals and counting those that are in bonding MO's and those that are in antibonding MO's. The equation for bond order is:

$$BO = \frac{(\text{no. of electrons in bonding MO's} - \text{no. of electrons in antibonding MO's})}{2}$$

(a) H_2 2 e⁻ $(\sigma_{1s})^2$ BO = 2/2 = 1

(b) C_2 8 e⁻ $(\sigma_{2s})^2 (\sigma^*_{2s})^2 (\pi_x)^2 (\pi_y)^2$

$$BO = \frac{(6-2)}{2} = 2$$

(c) N_2 10 e⁻ $(\sigma_{2s})^2 (\sigma^*_{2s})^2 (\pi_x)^2 (\pi_y)^2 (\sigma_p)^2$

$$BO = \frac{(8-2)}{2} = 3$$

(d) O_2 12 e⁻ $(\sigma_{2s})^2 (\sigma^*_{2s})^2 (\sigma_p)^2 (\pi_x)^2 (\pi_y)^2 (\pi^*_x)^1 (\pi^*_y)^1$

$$BO = \frac{(8-4)}{2} = 2$$

(e) F_2 14 e⁻ $(\sigma_{2s})^2 (\sigma^*_{2s})^2 (\sigma_p)^2 (\pi_x)^2 (\pi_y)^2 (\pi^*_x)^2 (\pi^*_y)^2$

$$BO = \frac{(8-6)}{2} = 1$$

4A-15 Calculate the bond order for the molecules. The molecule with the highest bond order is more stable.

O_2 12 e⁻ $(\sigma_{2s})^2 \ (\sigma^*_{2s})^2 \ (\sigma_p)^2 \ (\pi_x)^2 \ (\pi_y)^2 \ (\pi^*_x)^1 \ (\pi^*_y)^1$

BO = 2

O_2^{2-} 14 e⁻ $(\sigma_{2s})^2 \ (\sigma^*_{2s})^2 \ (\sigma_p)^2 \ (\pi_x)^2 \ (\pi_y)^2 \ (\pi^*_x)^2 \ (\pi^*_y)^2$

BO = 1

O_2 is more stable than O_2^{2-}.

4A-17 A molecule is paramagnetic if it contains unpaired electrons. The peroxide ion, O_2^{2-} has the electron configuration:
$(\sigma_{2s})^2 \ (\sigma^*_{2s})^2 \ (\sigma_p)^2 \ (\pi_x)^2 \ (\pi_y)^2 \ (\pi^*_x)^2 \ (\pi^*_y)^2$

All the electrons are paired. The peroxide ion is not paramagnetic.

4A-19 A molecule is paramagnetic if it contains unpaired electrons. It is diamagnetic when all electrons are paired.

(a)   HF   8 e⁻ $(\sigma_{s-p})^2$

Because the hydrogen and fluorine atoms are so different, the MO's for this molecule do not follow the general pattern. The σ_{s-p} MO is occupied by the 1s electron of hydrogen and one of the 2p electrons of fluorine. The other valence electrons are localized on the fluorine atom and do not participate in the MO bonding scheme.
All electrons are paired. HF is diamagnetic.

(b)   C O   10 e⁻ $(\sigma_{2s})^2 \ (\sigma^*_{2s})^2 \ (\sigma_p)^2 \ (\pi_x)^2 \ (\pi_y)^2$
All electrons are paired. CO is diamagnetic.

(c)   CN⁻   10 e⁻ $(\sigma_{2s})^2 \ (\sigma^*_{2s})^2 \ (\sigma_p)^2 \ (\pi_x)^2 \ (\pi_y)^2$
All electrons are paired. CN⁻ is diamagnetic.

(d)   N O   11 e⁻ $(\sigma_{2s})^2 \ (\sigma^*_{2s})^2 \ (\sigma_p)^2 \ (\pi_x)^2 \ (\pi_y)^2 \ (\pi^*_x)^1$
NO has an unpaired electron. It is paramagnetic.

(e)   NO⁺   10 e⁻ $(\sigma_{2s})^2 \ (\sigma^*_{2s})^2 \ (\sigma_p)^2 \ (\pi_x)^2 \ (\pi_y)^2$
All electrons are paired. NO⁺ is diamagnetic.

Chapter 5
Ionic and Metallic Bonds

5-1   The elements of the third row of the periodic table in order of decreasing metallic character are Na > Mg > Al > Si > P > S > Cl > Ar and in order of increasing nonmetallic character are Na < Mg < Al < Si < P < S < Cl < Ar.
   In general, elements decrease in metallic properties going from left to right across a row, and they increasingly take on non-metal characteristics moving from left to right across the same row. Along a diagonal band, elements such as B, Si, Ge, As, Sb, Te, and At are identified as semi-metals because they exhibit properties of both metallic and nonmetallic elements. In the third row of the periodic table the metals are Na, Mg, Al. Si is a semi-metal. P, S, Cl, and Ar are non-metals.

5-3   In order of increasing non-metallic character:
   (a)   Sr < Ge < Al < N
   (b)   Rb < K < Mg < Si
   (c)   Ge < As < P < N
   (d)   Al < B < N < F The correct sequence.

5-5   Activity is measured by having a low first ionization energy.
   (a) Ca   (b) Na   (c) K   (d) Mg

5-7   All the alkali metals have the same valence configuration of xs^1 and very small AVEE values. Thus it is easy for these atoms to form +1 ions.

5-9   Xenon

5-11   Since there is only one valence electron for all the elements in Group I, the AVEE is the same as first ionization energy. As one goes down the column the number of shells increases. As the number of shells increases the ionization energy of an electron from that shell decreases. So the ionization energy (and AVEE) will decrease from Li to Fr.

5-13   The second ionization energy for Li would be much higher than the first for two reasons. First, the removal of an electron from a +1 ion requires more energy than removing it from a neutral atom (see Coulomb's law). Second, the second ionization would remove an electron from a smaller shell than the first ionization. This means the second electron is removed from a smaller distance from the nucleus that also increases the ionization energy.

5-15 Be^{2+}: [He]
   Mg^{2+}: [Ne]
   Ca^{2+}: [Ar]
   Sr^{2+}: [Kr]
   Ba^{2+}: [Xe]
   Ra^{2+}: [Rn]

5-17   Na has a lower first ionization than Mg because they are filling electrons in the same shell, but Na has fewer protons in the nucleus.

5-19   As one goes down the column the number of shells increases. The more shells the larger the atomic size, therefore the atomic size increases from Be to Ra.

5-21   Aluminum

5-23   From B to Al the ionization energy and the AVEE decrease because the valence shell is larger for Al than for B. So it is easier to pull off these outer electrons.

5-25   Ga: $[Ar]4s^2 3d^{10} 4p^1$
Ga$^+$: $[Ar]4s^2 3d^{10}$
Ga^{+2}: $[Ar]4s^1 3d^{10}$
Ga^{+3}: $[Ar]3d^{10}$
The fourth ionization from Ga would be more difficult than the third, since the removal of an electron from a +2 ion requires more energy than removing it from a +3 ion (see Coulomb's law). Second, the fourth ionization would remove an electron from the filled 3d subshell. This means the fourth electron is removed from a smaller distance from the nucleus that also increases the ionization energy.

5-27   (a) As is in Group VA   5-8=-3   As^{3-}
(b) Te is in Group VIA   6-8=-2   Te^{2-}
(c) Se is in Group VIA   6-8=-2   Se^{2-}

5-29   The hydrogen of metal hydrides is a negative ion, H$^-$. The hydrogen atom in nonmetal hydrides is assigned a +1 oxidation state, H$^+$. In the nonmetal compounds of hydrogen, the hydrogen is covalently bound, which means that an electron pair is shared between the two bound atoms.

5-31   All the halogens have the same valence configuration of xp^5 and very large AVEE values. However, they also have high electron affinities. Thus it is easy for these atoms to attach an electron to close their electronic shell and form –1 ions.

5-33   Hydrogen has an AVEE that is almost in the middle of all main-group elements. When it reacts with an element with a low AVEE, it behaves like a halogen (NaH, sodium hydride). When it reacts with a high AVEE it behaves like an alkali metal (HBr, hydrogen bromide).

5-35   Zn: $[Ar]4s^2 3d^{10}$. The most likely change on a zinc ion would be +2, since the atom would lose the two 4s electrons and have a closed 3d shell.

5-37   Ni: $[Ar]4s^2 3d^8$
Ni^{2+}: $[Ar]\ 3d^8$

5-39   Fluorapatite, $Ca_5(PO_4)_3F$, contains five Ca^{2+} and three PO_4^{3-} ions, therefore it has a charge of: (5)(+2)+(3)(-3)=10-9=+1 after counting up the contribution of the Ca^{2+} and PO_4^{3-} ions. Since fluorapatite is a neutral molecule, the charge on the fluoride ion must be –1.

5-41   (a)$Mg(NO_3)_2$ (b)$Fe_2(SO_4)_3$ (c)Na_2CO_3

5-43   Magnesium nitride, Mg_3N_2 contains the Mg^{2+} ion and the N^{3-} ion. From the charge on the nitride ion, and the fact that the sodium ion is Na$^+$ and the aluminum ion is Al^{3+}, the formula for sodium nitride is Na_3N and the formula for aluminum nitride is AlN.

5-45 $[Co(NO_2)_x]^{3-}$ The sum of the charges on the ions in the formula must equal -3, the charge on the complex ion. +3 + x(-1) = -3, therefore x = 6

5-47 (a) Na_2O, NaH (b) MgO, MgH_2 (c) Al_2O_3, AlH_3 (d) SiO_2, SiH_4
(e) P_2O_5, PH_3
The element best matching the suggested compositions is magnesium, b.

5-49 Main-group elements matching the reactivities described are to be found in Group IIIA.

5-51 The very high third ionization energy for magnesium (7733 kJ) means that Mg will not readily form a +3 ion.

5-53 The product of the reaction of aluminum (lose 3 valence electrons) with nitrogen (gain 3 valence electrons) should have the formula AlN.

5-55 Gallium of Group IIIA combining with arsenic of Group VA should give a product with the formula GaAs.

5-57 (a) $Ca(s) + H_2(g) \rightarrow CaH_2(s)$
(b) $2\ Ca(s) + O_2(g) \rightarrow 2\ CaO(s)$
(c) $8\ Ca(s) + S_8(s) \rightarrow 8\ CaS(s)$
(d) $Ca(s) + F_2(g) \rightarrow CaF_2(s)$
(e) $3\ Ca(s) + N_2(g) \rightarrow Ca_3N_2(s)$
(f) $6\ Ca(s) + P_4(s) \rightarrow 2\ Ca_3P_2(s)$

5-59 **Superoxides** are compounds that form when a very reactive alkali metal, such as potassium, reacts with a 1:1 ratio, with the O_2 molecule to form KO_2.

5-61 The Lewis structures for
Rb   Rb$^+$ O^{2-}

$\cdot$ Rb   Rb   :O:

For rubidium oxide and rubidium superoxide the structures are:

Rb :O: Rb   :O——O: Rb

5-63 Ionic compounds are held together by electrostatic or coulombic forces that hold together oppositely charged species.

5-65

:Cl: Ba: Cl: Na: S : Na   Ca: S :

5-67   a) The attractive force depends on the magnitudes of the q values and the distance between the two charges.
b) At a fixed distance if the q's are bigger the attractive force is bigger, therefore the +2, –2 pair would have the stronger attraction.
c) The attraction would be greater at smaller distance since the distance squared is in the denominator.
d) NaCl would have the stronger attractive forces since the Na^+ ion is smaller than the K^+ ion making the distance between the ions smaller.

e) For Li_2O the attractive force can be given as $F = \dfrac{(+1)(-2)}{(0.068 + 0.14)^2} = -46.2$

whereas for CaO it is $F = \dfrac{(+2)(-2)}{(0.099 + 0.14)^2} = -70.0$.

So even though the Ca^{+2} ion is larger than the Li^+ ion, the different charge has a larger effect in this case.

5-69

5-71   Ionic and covalent bonds are similar in that the electrons of the bonded atoms are localized. That is, the electrons reside either on a particular ion (ionic bonding) or are shared by two different atoms (covalent bonding).

5-73   A metallic bond is one in which a group of positive metal ions are held together by a mobile sea of negative electrons.

5-75   Atoms that engage in metallic bonding are typically large and have very low values of AVEE as a result. Furthermore they don't have enough valence electrons to possibly share enough electrons to fill their valence shell, therefore losing them to form a metallic bond takes very little energy.

5-77   The difference between ionic and covalent bonds rests in the extent to which the electrons are shared by the atoms that form the bond. In an ionic bond there is a transfer of electrons from one atom to another forming a positive ion and a negative ion. In a covalent bond the two atoms share a pair of electrons. Chemists believe ionic and covalent are the extremes of a continuum because there is a large group of molecules that are not purely ionic nor purely covalent. In these molecules, the electrons in a bond are not shared equally, so there is some separation of charge.

5-79   The electronegativity difference is the smallest for O and Se. Response (e).

5-81   From the bond type triangle
(a) CaH_2 ionic
(b) BrF_3 polar covalent
(c) NF_3 polar covalent
(d) $SiCl_4$ polar covalent
(e) AsH_3 covalent/semimetallic
(f) $MgZn_2$ metallic

46

5-83 (a), (b), (e)

5-85 (e)

5-87 Using the guide that ionic substances generally result when metals combine with non-metals.
      (a) IF_3 covalent
      (b) $SiCl_4$ covalent
      (c) BF_3 covalent
      (d) Na_2S ionic

Using a bond type triangle as a guide.
      (a)   covalent
      (b)   covalent
      (c)   covalent
      (d)   ionic

5-89 No, the compounds with low ΔEN could be metallic.

5-91
(a) Hg=1.76	S=2.59	$\overline{EN}$ =2.18 ΔEN=0.83	Covalent
(b) Ga=1.76	Sb=1.98	$\overline{EN}$=1.87 ΔEN=0.22	Semi-metal
(c) Li=0.91	N=3.07	$\overline{EN}$=1.99 ΔEN=2.16	Ionic
(d) Na=0.87	Br=2.69	$\overline{EN}$=1.78 ΔEN=1.82	Ionic
(e) Sn=1.82	Br=2.69	$\overline{EN}$=2.26 ΔEN=0.87	Covalent
(f) Na=0.87	P=2.25	$\overline{EN}$=1.56 ΔEN=1.38	Ionic/Metallic
(g) In=1.66	P=2.25	$\overline{EN}$=1.96 ΔEN=0.59	Covalent with some semi-metal character
(h) In=1.66	N=3.07	$\overline{EN}$=2.37 ΔEN=1.41	Ionic/Covalent
(i) Te=2.16	O=3.61	$\overline{EN}$=2.88 ΔEN=1.45	Covalent

5-93 The bonding between Mn and Al would be semimetallic because the difference in electronegativity is very small. Given that it has semimetallic bonding this compound is a semiconductor.

5-95 a) BBr_3 covalent; Li_3P metallic; SrF_2 ionic; $SrZn_5$ metallic; Cd_3N_2 ionic.
    b)   i) SrF_2, Cd_3N_2
           ii) BBr_3
           iii) $SrZn_5$, Li_3P

5-97 Mg_3N_2 is an ionic compound since ΔEN is 1.78 and the $\overline{EN}$ is 2.18.

$BaSi_2$ is a metallic compound since ΔEN is 1.04 and the $\overline{EN}$ is 1.4.

B_2O_3 is a covalent compound since ΔEN is 1.56 and the $\overline{EN}$ is 2.83.

a) Covalent molecules are insulators therefore B_2O_3 is an insulator. Ionic compounds can also be insulators provided they are solid. Therefore Mg_3N_2 is also an insulator.
b) This describes an ionic compound, Mg_3N_2.
c) Metallic compounds conduct electricity in the solid state, $BaSi_2$.

5-99   (a) K_3P $3(+1) + x = 0$ $x = -3$, P^{3-}

       (b) Na_3PO_4 $1(+3) + x + 4(-2) = 0$ $x = +5$, P^{5+}

       (c) PO_3^{3-} $x + 3(-2) = -3$ $x = +3$, P^{3+}

       (d) P_2Cl_4 $2x + 4(-1) = 0$ $x = +2$ P^{2+}

5-101 $NaAl(OH)_2CO_3$ $1(+1) + x + 2(-1) + 1(-2) = 0$ $x = +3$, Al^{3+}

5-103   Taking oxygen as O^{2-}, the oxidation number assigned to the chlorine atom will be

       (a) 0   (b) -1   (c) +1   (d) +3   (e) +5   (f) +7

5-105   Assigning the oxidation number of -1 to fluorine and -2 to oxygen, the assigned oxidation number for xenon in the compounds will be

       (a)+2 (b)+4 (c)+4   (d)+6   (e)+6   (f)+6   (g)+8   (h)+8

       Xenon exhibits a positive oxidation number from +2 to +8 and includes only even numbers.

5-107   (a) 0   (b)-2   (c) -2   (d) +4   (e) +6   (f) +4   (g) +6   (h) +4

       (i) +6   (j)+4   (k) +6

       Sulfur has an oxidation number of -2 in H_2S and ZnS.

       Sulfur has an oxidation number of +4 in SF_4, SO_2, SO_3^{2-}, and H_2SO_3.

       Sulfur has an oxidation number of +6 in SF_6, SO_3, SO_4^{2-}, and H_2SO_4.

       Sulfur shows even oxidation numbers from -2 to +6.

5-109   +2)   TiO

       +3) Ti_2O_3, Ti_2S_3, $TiCl_3$

       +4) TiO_2, $TiCl_4$, K_2TiO_3, H_2TiCl_6, $Ti(SO_4)_2$

       Titanium has oxidation numbers from +2 to +4.

5-111   (a) C_{CH3} =-3 $C_{central} = +1$ H = +1   Br = -1

       (b)   C = -2   H = +1   O = -2

       (c) C_{CH3}= -3 $C_{central} = +2$ H = +1   O = -2

       (d) C_{CH3} = -3 $C_{central} = 0$ H = +1   O = -2

5-113   (a) C_{CH3} = -3 $C_{central} = +3$ H = +1   O = -2   Cl = -1

       (b)   C = -2   H = +1   N = -3

       (c) C_{CH3} = -3 $C_{C=O} = +3$ $C_{CH2} = -1$ O = -2   H = +1

       (d)   C = -3   H = +1

       (e) $C_{CH2} = -2$ $C_{CH} = -1$ $C_{CH3} = -3$ H = +1

       (f)   C = -1   H = +1

5-115   The partial charge on the H in HCl is 0.11. The partial charge on the Cl is –0.11. However, the Lewis structure for HCl indicates that each of the atoms is assigned a formal charge of zero. In assigning oxidation numbers to HCl the shared electrons are assigned to the more electronegative element, which would be the Cl. This would then have an oxidation number of –1 and the hydrogen has an oxidation number of +1.

5-117 (a) This is an oxidation-reduction reaction. Magnesium is oxidized from 0 in Mg(s) to +2 in $MgCl_2$. Hydrogen is reduced from +1 in HCl to 0 in H_2.
(b) This is an oxidation-reduction reaction. Iodine is oxidized from 0 in I_2 to +3 in ICl_3. Chlorine is reduced from 0 in Cl_2 to -1 in ICl_3.
(c) This is not an oxidation-reduction reaction.
(d) This is an oxidation-reduction reaction. Sodium is oxidized from 0 in Na(s) to +1 in NaOH. Hydrogen is reduced from +1 in H_2O to 0 in H_2.

5-119 In the reaction $4 C(s) + S_8(l) \rightarrow 4 CS_2(l)$, the oxidation state of carbon increases from 0 in C(s) to +4 in $CS_2(l)$. Carbon is therefore oxidized. The sulfur changes from an oxidation state of 0 in S_8 to -2 in CS_2. Sulfur is therefore reduced.

5-121 (a) The name phosphorus pentoxide for P_2O_5 does not indicate the number of phosphorus atoms present. Diphosphorus pentoxide would be a better name.
(b) The name iron oxide for Fe_2O_3 does not indicate the charge on the iron. The charge needs to be specified so that you can distinguish whether you are discussing FeO or Fe_2O_3. A better name is iron (III) oxide.
(c) The name chlorine monoxide for Cl_2O does not indicate the number of chlorine atoms. A better name is dichlorine monoxide.
(d) The name copper bromide for $CuBr_2$ does not indicate the oxidation number of the copper atom. A better name is copper(II) bromide.

5-123 (a) P_4S_3 (b) SiO_2 (c) CS_2 (d) CCl_4 (e) PF_5

5-125 (a) $SnCl_2$ (b) $Hg(NO_3)_2$ (c) SnS_2 (d) Cr_2O_3 (e) Fe_3P_2

5-127 (a) $Co(NO_3)_3$ (b) $Fe_2(SO_4)_3$ (c) $AuCl_3$ (d) MnO_2 (e) WCl_6

5-129 (a)   aluminum chloride
      (b)   sodium nitride
      (c)   calcium phosphide
      (d)   lithium sulfide
      (e)   magnesium oxide

5-131 (a)   antimony(III) sulfide
      (b)   tin(II) chloride
      (c)   sulfur tetrafluoride
      (d)   strontium bromide
      (e)   silicon tetrachloride

5-133 (a) H_2CO_3
      (b)   HCN
      (c) H_3BO_3
      (d) H_3PO_3
      (e) HNO_2

5-135 thiosulfate ion, SSO_3^{2-} or $S_2O_3^{2-}$

5-137 (a) calcite, calcium carbonate
(b) barite, barium sulfate

5-139

Compound	ΔEN	$\overline{EN}$	Bond Triangle Classification
FeO	1.94	2.64	Covalent
Fe_2O_3	1.94	2.64	Covalent
$FeCl_2$	1.20	2.27	Covalent
$FeCl_3$	1.20	2.27	Covalent

All of the above compounds are classified as covalent according to bond type. They all lie however, close to the dividing line between ionic and covalent bonding.

5-141

	I	Br
Partial Charge	+0.065	-0.065
Formal Charge	0	0
Oxidation Number	+1	-1

In the partial charge calculation, the more electronegative (EN) atom, Br, is partially negative and the less EN atom, I, is partially positive. Remember EN decreases down a column. When assigning oxidation numbers, the less EN atom is given a positive oxidation number. Since IBr is neutral, the oxidation numbers must add to zero. Thus, Br has an oxidation number of -1 and I of +1. The formal charge for each atom is zero. The partial charge representation is the most accurate representation. The formal charge calculation is useful when drawing Lewis structures. The sum of the formal charges must equal the overall charge of the molecule or ion. Oxidation numbers are useful when describing oxidation-reduction reactions.

5-143 There is an ionic bond between the Ba^{2+} two NO_3^- ions. Within the nitrate ion there are covalent bonds between N and O. This compound is different because it contains the polyatomic ion NO_3^-. The Lewis structure for $Ba(NO_3)_2$ is shown below.

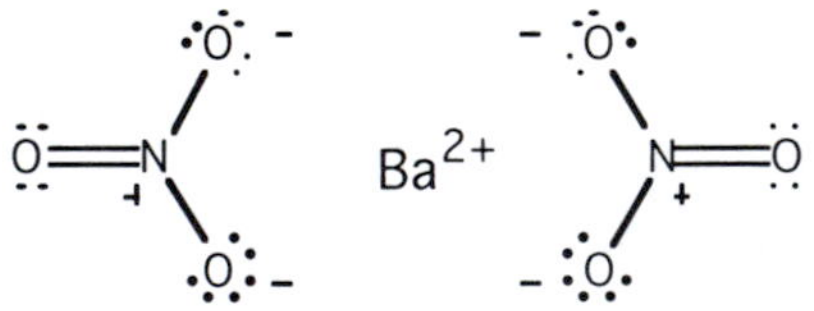

5-145 (a) C_3H_6

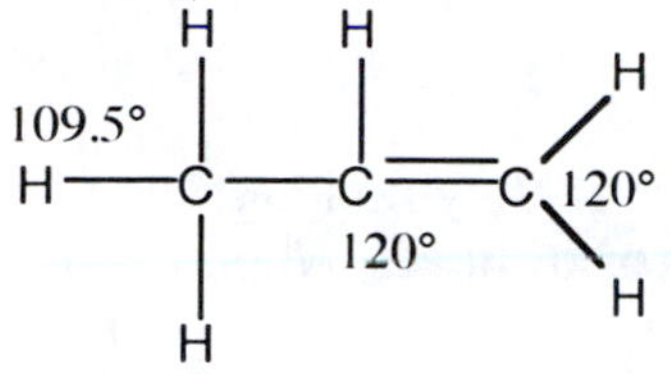

C_3H_3N

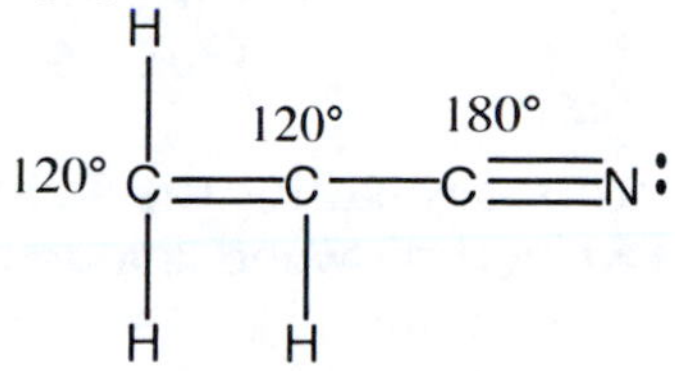

(b) (i) False, the nitrogen in the NO is reduced from a state of +2 to 0. The oxygen doesn't change oxidation state.

(ii) False, the carbon on the C_3H_6 is oxidized from a –3 to a +3 state.

(iii) False, in an oxidation-reduction reaction, at least one of the reactants must be oxidized and another must be reduced. Both reactants cannot be reduced. See the answers for (i) and (ii) for further explanations

(iv) True.

5-147 (a) $2\ PbS + 3\ O_2 \rightarrow 2\ PbO + 2\ SO_2$.

(b) It is an oxidation-reduction reaction. The O_2 is reduced and the S is oxidized.

5-149 MgO. High ΔEN indicates that it is an ionic compound that does not conduct electricity well. Ionic bonds are also very strong.

Chapter 6
Gases

6-1   Metals are good conductors of heat for the same reason that they conduct electricity well. Because of its covalent structure, wood is a poor conductor of heat. When both materials are at the same temperature, below skin temperature, the metal will feel colder because it is carrying the heat away from your skin faster than the wood.

6-3   The **temperature** of a system is a measure of the average kinetic energy of all the particles in the system.

6-5   Water freezes at 32°F, 0°C or 273.15 K.

6-7   Kinetic energy can be thought of as the energy of motion. The faster something is moving, the greater its kinetic energy.

6-9   Many of the properties of gases are independent of the chemical composition of the gas and can be explained by the kinetic molecular theory model.

6-11   Elements of the upper right hand portion of the periodic table and covalent compounds derived from combinations of these elements are most likely to be gases at room temperature. Of those covalent compounds, those with the smaller masses are more likely to be gases. Gases: Ar, CO, CH_4, Cl_2

6-13   Low atomic weight elements that are non-metals are typically gases. Also low molecular weight covalent molecules are commonly gases.

6-15   The volume of a mole of liquid water at 25°C and 1 atm will be about 18 mL (the molecular mass of water is 18 g/mole and the density of water at 25°C is about 1 g/mL). However, the volume of a mole of gaseous water is about 22 L (see below).

6-17   The mass of the cylinder will increase when helium is added to the cylinder. A vacuum has no mass. Helium has a mass of 4.00 grams/mole.

6-19   The pressure of the wind will be the same on the sign and the billboard. However, the billboard has a larger area, therefore it must be feeling a greater force against it than the stop sign, which has a smaller area.

6-21 $745.8 \text{ mm Hg} \times \dfrac{1 \text{ atm}}{760 \text{ mm Hg}} = 0.9813 \text{ atm}$

$0.9813 \text{ atm} \times \dfrac{1.013 \times 10^5 \text{ Pa}}{1 \text{ atm}} = 9.941 \times 10^4 \text{ Pa}$

6-23 $760 \text{ mm Hg} \times \dfrac{1 \text{ cm}}{10 \text{ mm}} \times \dfrac{1 \text{ in}}{2.54 \text{ cm}} = 29.9 \text{ in Hg}$

$1 \text{ atm} = 1.01 \times 10^5 \text{ Pa}$

6-25 $2 \times 10^{-3} \text{ mm Hg} \times \dfrac{1 \text{ atm}}{760 \text{ mm Hg}} = 3 \times 10^{-6} \text{ atm}$

6-27 $P_1V_1 = P_2V_2$ T=constant
P_1=742.3 mm Hg V_1=425 ml
P_2=? V_2=975 ml

$$P_2 = \frac{P_1V_1}{V_2} = \frac{742.3 \text{ mm Hg} \times 425 \text{ ml } O_2}{975 \text{ ml } O_2} = 324 \text{ mm Hg}$$

6-29 $P_1V_1 = P_2V_2$ T=constant
P_1= 1.0 atm V_1=2000 L
P_2=150 atm V_2=? L

$$V_2 = \frac{P_1V_1}{P_2} = \frac{1.0 \text{ atm} \times 2000 \text{ L}}{150 \text{ atm}} = 13 \text{ L}$$

6-31 $\dfrac{P_1}{P_2} = \dfrac{T_1}{T_2}$

P_1=5.00 atm P_2=?
T_1=21°C=294 K T_2=38°C=311 K

$$P_2 = \frac{P_1T_2}{T_1} = \frac{5.00 \text{ atm} \times 311 \text{ K}}{294 \text{ K}} = 5.29 \text{ atm}$$

6-33 The fraction of the total pressure due to the pressure of each gas will remain unchanged as the temperature increases.

6-35 $\dfrac{V_1}{V_2} = \dfrac{T_1}{T_2}$

T_1=21°C=294 K T_2=0°=274 K
V_1=0.357 L V_2=?

$$V_2 = \frac{V_1T_2}{T_1} = \frac{0.357 \text{ L} \times 273 \text{ K}}{294 \text{ K}} = 0.332 \text{ L}$$

6-37 $2 \text{ NH}_3(g) \rightarrow N_2(g) + 3 \text{ H}_2(g)$

$$1.38 \text{ L NH}_3 \times \frac{1 \text{ mol } N_2}{2 \text{ mol NH}_3} = 0.690 \text{ L } N_2$$

$$1.38 \text{ L NH}_3 \times \frac{3 \text{ H}_2}{2 \text{ NH}_3} = 2.07 \text{ L } H_2 \qquad \frac{\text{volume } H_2}{\text{volume } N_2} = \frac{2.07 \text{ L}}{0.69 \text{ L}} = 3:1$$

6-39 $2 \text{ C}_2\text{H}_2(g) + 5 \text{ O}_2(g) \rightarrow 4 \text{ CO}_2(g) + 2 \text{ H}_2O(g)$

15.0 L of C_2H_2 require 37.5 L O_2 to react. Therefore, O_2 is the limiting reagent.

$$15.0 \text{ L } O_2 \times \frac{4 \text{ mol CO}_2}{5 \text{ mol } O_2} + 15.0 \text{ L } O_2 \times \frac{2 \text{ mol H}_2O}{5 \text{ mol } O_2} = 12.0 \text{ L CO}_2 + 6.0 \text{ L H}_2O = 18.0 \text{ L Tot Vol}$$

6-41 At the same temperature and pressure equal volumes of gases will contain equal moles of gas. If n is the same for dry air and water vapor, then dry air will weigh more at 29.0 g/mol than water at 18.0 g/mol.

6-43 Look at the ratio of the volumes of the gases:
2.36 L Nitrous oxide(g)→2.36 L N_2(g)+1.18 L O_2(g)
or 2 Nitrous oxide → 2 N_2+1 O_2
The formula must be N_2O.

6-45 PV=nRT

(a)

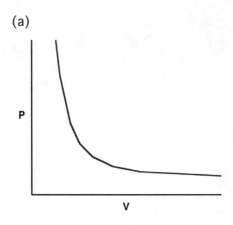

(b)

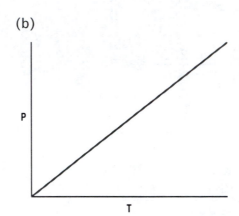

(c)

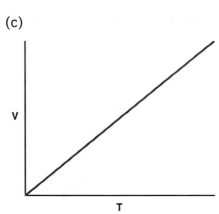

(d)

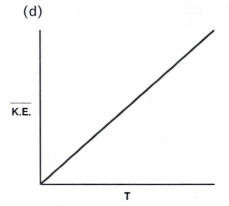

(e)

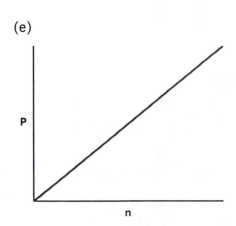

6-47 (a) P=constant $\frac{V}{T}=nR$, n will increase only if V increases more than T. This statement is not always true for an ideal gas.

(b) $\dfrac{P_1V_1}{T_1}=\dfrac{P_2V_2}{T_2}$ $\qquad V_1\left(\dfrac{P_1}{P_2}\right)\left(\dfrac{T_2}{T_1}\right)=V_2$ where $P_2>P_1$ and $T_2<T_1$

$\dfrac{P_1}{P_2}$ is less than 1 and $\dfrac{T_2}{T_1}$ is less than 1, therefore V_2 must always be less

than V_1. The statement is always true for an ideal gas.

(c) $\dfrac{P_1V_1}{n_1}=\dfrac{P_2V_2}{n_2}$ $\qquad P_1\left(\dfrac{V_1n_2}{V_2n_1}\right)\left(\dfrac{T_2}{T_1}\right)=P_2$ where $V_1>V_2$ and $n_1>n_2$

$\dfrac{V_1}{V_2}$ is greater than 1, $\dfrac{n_2}{n_1}$ is less than 1.

Only if the change in V were greater than the corresponding change in n would the pressure need to decrease. The statement is not always true for an ideal gas.

6-49 Starting from the accepted value of R in the accepted units, $\dfrac{\text{L atm}}{\text{mol}\cdot\text{K}}$, we convert the liters to cubic centimeters. One cubic centimeter = 1 mL. Thus 1000 cc=1 L.

$$R = 0.08206\frac{\text{L atm}}{\text{mol}\cdot\text{K}}\cdot\frac{1000\ \text{cc}}{1\ \text{L}}=82.06\frac{\text{cc atm}}{\text{mol}\cdot\text{K}}$$

6-51 If there are equal numbers of gas particles, then there will be equal pressures of each.

$P_{He}=P_{O_2}=P_{N_2}$ will each be 1/3 of the total pressure=2.5 atm

6-53 $P_{CO_2}=\dfrac{nRT}{V}=\dfrac{0.450\ \text{g}\times\dfrac{1\ \text{mol CO}_2}{44.009\ \text{g}}\times\dfrac{0.08206\ \text{L}\cdot\text{atm}}{\text{mol}\cdot\text{K}}\times300\ \text{K}}{1\ \text{L}}=0.252\ \text{atm}$

$P_{Total}=P_{CO}+P_{CO_2}=0.200+0.252=0.452\ \text{atm}$

6-55 $\dfrac{\$0.50}{100\ \text{ft}^3}\times\dfrac{1\ \text{ft}^3}{28.316\ \text{L}}\times\dfrac{22.4\ \text{L}}{1\ \text{mol N}_2}\times\dfrac{1\ \text{mol N}_2}{28.014\ \text{g}}=\dfrac{\$0.00014}{\text{g N}_2}$

6-57 PV=nRT

$T=\dfrac{PV}{nR}=\dfrac{740\ \text{mm Hg}\times\dfrac{1\ \text{atm}}{760\ \text{mm Hg}}\times1\ \text{L}}{1.50\ \text{g O}_2\times\dfrac{1\ \text{mol O}_2}{31.9988\ \text{g}}\times0.08206\dfrac{\text{L}\cdot\text{atm}}{\text{mol}\cdot\text{K}}}=253\ \text{K}$

6-59 $PV=nRT$ $\qquad P=\dfrac{nRT}{V}=\dfrac{gRT}{MW\,V}=\dfrac{dRT}{MW}$ where $d=$ density.

$$P=1.118\dfrac{g}{cm^3}\times 1\;cm^3\times\dfrac{\dfrac{1\;mol\;O_2}{31.9988\;g}\times 0.08206\dfrac{L\cdot atm}{mol\cdot K}\times 273\;K}{0.250\;L}=3.13\;atm$$

6-61 $PV=nRT$ $\qquad \dfrac{n}{V}=\dfrac{P}{RT}=\dfrac{1\,atm\cdot mol\cdot K}{0.08206\;L\cdot atm\times 313\;K}=0.0389\dfrac{mol}{L}$

1 mol $CH_2Cl_2=84.933$ g therefore $\dfrac{0.0389\;mol\times 84.933\dfrac{g}{mol}}{L}=3.30\dfrac{g}{L}$ density of CH_2Cl_2 gas.

In the liquid phase, CH_2Cl_2 has a density of $1.336\dfrac{g}{cm^3}\times\dfrac{1\;cm^3}{1\;ml}\times\dfrac{1000\;ml}{1\;L}=1.336\times 10^3\dfrac{g}{L}$

The density of the liquid is $\cong 400$ times greater than the density of the gas.

6-63 1 mol ideal gas occupies 22.4 L at STP

$$\dfrac{1\;mol\;He}{22.4\;L}=\dfrac{4.0026\;g}{22.4\;L}=0.179\dfrac{g}{L}=\text{density Helium at STP}$$

The density of air is 1.29 g/L at STP

The lift associated with a balloon is the difference in weight between the displaced gas and that gas doing the displacing.

$$\dfrac{1.29\;g\;air-0.179\;g\;He}{1\;L}=\dfrac{1.11\,g}{L}\times\dfrac{28.316\;L}{ft^3}\times\dfrac{1\;lb}{453.6\;g}=0.0694\dfrac{lb}{ft^3}\text{lift, Therefore the}$$

balloon can lift a weight of 0.07 lbs.

6-65 $3.7493\dfrac{g}{L}\times\dfrac{22.4\;L}{mol}=84.0\dfrac{g}{mol}$; The gas is krypton.

6-67 $\dfrac{V_1}{T_1}=\dfrac{V_2}{T_2}$ $\qquad V_2=\dfrac{V_1T_2}{T_1}=\dfrac{(5.0\;L)(78\;K)}{(298\;K)}=1.3\;L$

6-69 $PV=nRT$ $\qquad P=\dfrac{nRT}{V}=\dfrac{4.80\;g\;O_3\times\dfrac{1\;mol\;O_3}{47.997\;g}\times\dfrac{0.08206\;L\cdot atm}{mol\cdot K}\times 298\;K}{2.45\;L}=0.998\;atm$

$2\;O_3(g)\to 3\;O_2$

$$\dfrac{P_1V_1}{n_1}=\dfrac{P_2V_2}{n_2}\qquad P_2=\dfrac{P_1n_2}{n_1}=\dfrac{(0.998\;atm)(3\;mol\;O_2)}{(2\;mol\;O_3)}=1.50\;atm$$

6-71 $T_2 = \dfrac{P_2 V_2 T_1}{P_1 V_1} = \dfrac{(0.978 \text{ atm})(10.0 \text{ L})(298 \text{ K})}{(2.5 \text{ atm})(5.0 \text{ L})} = 2.3 \times 10^2 \text{ K}$

6-73   If the gas molecules in a balloon were in a state of constant motion and the motion was not random, the balloon would not appear smoothly round. It would be elongated in the direction of the predominating motion and contracted in the other directions.

6-75   The pressure of a gas results from collisions between the gas particles and the walls of the container. Any increase in the number of gas particles in the container increases the number of collisions with the walls and therefore the pressure of the gas.
The average kinetic energy of the gas particles becomes larger as the gas becomes warmer. Since the mass of the gas particles is constant, this means that the average velocity of the particles must increase. The faster the particles are moving when they hit the wall, the greater the force they exert on the wall. Since the force per collision becomes larger as the temperature increases, the pressure of the gas must increase as well.
If we compress a gas without changing its temperature, the average kinetic energy of the gas particles stays the same. There is no change in the speed with which the particles move, but the container is smaller. Thus the particles travel from one end of the container to the other in a shorter period of time. This means that they hit the walls more often. Any increase in the number of collisions with the walls must lead to an increase in the pressure exerted by the gas. Thus the pressure exerted by a gas becomes larger as the volume of the gas becomes smaller.

6-77   (a) 7.0×10^{-21} J/molecule. The temperature determines the average kinetic energy of the molecules. If the NH_3 molecules are at 50°C and have an average kinetic energy of 7.0×10^{-21} J/molecule, then O_2 which is also at 50°C will have the same average kinetic energy.
(b) NH_3. Even though both molecules have the same average kinetic energy, that energy is given by $\frac{1}{2}mv^2$. Since a molecule of O_2 is heavier than NH_3 it will have a lower velocity.

(c) No, we would also need the volume and the number of molecules present to determine the pressure $PV=nRT$.
(d) Since the average kinetic energy is proportional to the temperature, we would need to increase the temperature.
(e) If the temperature is kept the same, then the average kinetic energy is the same, but the pressure would be reduced.
(f)   (1) Fix n and V, double the pressure.
       (2) Fix T and V, double the number of molecules present.
       (3) Fix n and T, cut the volume in half.

6-79   V is a constant so using PV=nRT,

$$\text{for A}\quad PV = 1.0\,mol \times \frac{0.08206\ L \cdot atm}{mol \cdot K} \times 293.15\ K = 24.1\,atm \cdot V$$

$$\text{for B}\quad PV = 1.0\,mol \times \frac{0.08206\ L \cdot atm}{mol \cdot K} \times 308.15\ K = 25.3\,atm \cdot V$$

$$\text{for C}\quad PV = 2.0\,mol \times \frac{0.08206\ L \cdot atm}{mol \cdot K} \times 293.15\ K = 48.1\,atm \cdot V$$

(a) From the above calculations, C.
(b) Both A and C have the lowest temperature and hence the lowest average kinetic energy.
(c) B has a higher temperature therefore a higher average kinetic energy. Since all flasks have atoms of the same mass, the one with the highest kinetic energy will also have the highest velocity.
(d) C. Pressure is determined by the number of collisions with the walls. The more collisions, the higher the pressure.

6-81   (a) i) increase   ii) remain the same   iii) remain the same   iv) decrease
(b) remain the same
(c) i) increase   ii) remain the same (the changes counter balance)   iii) decrease
(d) The flask with He has more molecules since the pressure is higher while the temperature and volume are the same.

6-83   The **effusion** of a gas is the rate at which it will escape though a pinhole into a vacuum.

6-85   The rate of diffusion is inversely related to the molecular mass.
   (a)   Ar   39.95 g/mol   (b) Cl_2 70.91 g/mol
   (c) CF_2Cl_2 120.91 g/mol   (d) SO_2 64.06 g/mol
   (e) SF_6 146.05 g/mol
   In order of increasing rate of diffusion: $SF_6 < CF_2Cl_2 < Cl_2 < SO_2 < Ar$

6-87   Immediately after the valve is opened, the weight of the flask containing the hydrogen will decrease for two main reasons. First, since the rate of diffusion is inversely proportional to the square root of molecular mass, hydrogen will diffuse four times faster than oxygen. Secondly, the initial pressure of hydrogen will be 16 times greater than that of oxygen. Eventually, however, the weight of the two flasks will equalize as both gases spread evenly throughout both containers.

6-89   The heavier gas will effuse through the pinhole more slowly. Nitric oxide must be NO (the less massive gas) and nitrous oxide must be N_2O (the more massive gas).

6-91
$$\frac{\text{time unknown gas}}{\text{time } O_2} = \frac{60.0\ s}{84.9\ s} = \sqrt{\frac{MW_x}{31.9988}}$$

$$31.9988 \times \frac{(60.0\ s)^2}{(84.9\ s)^2} = MW_x = 16.0\,\frac{g}{mol}$$

6-93
$$\frac{\text{Rate Hg}}{\text{Rate Rn}} = \frac{1.082}{1} = \sqrt{\frac{x}{200.59\ g}}$$
$(200.59\ g)(1.082)^2 = x = 234.2\ g$ MW Radon $= 234.8\ g/mol$

6-95 $Mg(s)+2HCl(aq) \rightarrow Mg^{2+}(aq)+2\,Cl^-(aq)+H_2(g)$

$$\text{mol } H_2 = n = \frac{PV}{RT} = \frac{1.00\text{ atm} \times 0.500\text{ L}}{0.08206\dfrac{L\cdot atm}{mol\cdot K} \times 273\text{ K}} = 0.0223\text{ mol } H_2$$

$$0.0223\text{ mol } H_2 \times \frac{1\text{ mol Mg}}{1\text{ mol } H_2} \times \frac{24.305\text{ g Mg}}{1\text{ mol Mg}} = 0.542\text{ g Mg}$$

6-97 $CaCO_3(s) \rightarrow CaO(s) + CO_2(g)$

$$150\text{ kg CaCO}_3 \times \frac{1000\text{ g}}{1\text{ kg}} \times \frac{1\text{ mol CaCO}_3}{100.086\text{ g CaCO}_3} \times \frac{1\text{ mol CO}_2}{1\text{ mol CaCO}_3} = 1.50 \times 10^3\text{ mol CO}_2$$

$$V = \frac{nRT}{V} = \frac{1.50 \times 10^3 \times 0.08206\dfrac{L\,atm}{mol\,K} \times 296\text{ K}}{\dfrac{756}{760}\text{ atm}} = 3.66 \times 10^4\text{ L CO}_2$$

6-99 $$10.0\text{ g CaCO}_3 \times \frac{1\text{ mol CaCO}_3}{100.086\text{ g}} \times \frac{1\text{ mol CO}_2}{1\text{ mol CaCO}_3} = 0.0999\text{ mol CO}_2$$

$$V = \frac{nRT}{P} = \frac{0.0999\text{ mol CO}_2 \times 0.08206\text{ L}\cdot atm \times 296\text{ K}}{0.991\text{ atm mol}\cdot K} = 2.45\text{ L CO}_2$$

6-101 (a), (b), and (d) are correct.

6-103 $$\frac{g}{V} = \text{density} = \frac{P(MW)}{RT},\ MW = \frac{(\text{density})RT}{P} = \frac{5.86\text{ g} \times 0.08206\text{ L atm} \times 273\text{ K}}{1.00\text{ atm L mol K}}$$

=131 g/mol. The gas is xenon.

6-105 During a rainstorm, gaseous water molecules are condensing to form rain droplets. Thus, the total atmospheric pressure decreases.

6-107 $N_2H_4(g) \rightarrow N_2(g) + 2\,H_2(g)$
From the stoichiometry of the reaction, one mole of gaseous reactant produces three moles of gaseous products. The final pressure will be three times the initial pressure.

6-109 (a) Kinetic energy would remain unchanged, but the frequency of collisions would decrease.
(b) Kinetic energy would decrease as well as the frequency of collisions.
(c) Kinetic energy and frequency of collisions would both increase.
(d) Kinetic energy would remain unchanged, but the frequency of collisions would increase.

6-111   PV=nRT

$$n=\frac{PV}{RT}=\frac{\dfrac{756}{760}\,atm\times0.2755L}{0.08206\dfrac{L\cdot atm}{mol\cdot K}\times373\ K}=8.95\times10^{-3}\ mol\ acetone$$

$$MW\ acetone=\frac{0.520\ g}{8.95\times10^{-3}\ mol}=58.1\frac{g}{mol}$$

6-113   Find empirical formula: $85.63\ g\ C\times\dfrac{1\ mol\ C}{12.011g}=7.129\ mol\ C$

$$14.37\ g\ H\times\frac{1\ mol\ H}{1.0079\ g}=14.26\ mol\ H$$

Empirical formula is CH_2.

$0.45\ L\ (CH_2)_n + excess\ O_2 \rightarrow 1.35\ L\ CO_2 + 1.35\ L\ H_2O$

at the same T and P, mol amounts are proportional to volumes.

$L\ (CH_2)_n + excess\ O_2 \rightarrow 3\ L\ CO_2 + 3\ L\ H_2O$

for which n=3 and the formula for cyclopropane is C_3H_6.

Chapter 6 Special Topics

6A-1 Since real gas molecules are attracted to each other, they are pulled closer together and their volume will be smaller than that of an ideal gas at the same temperature and pressure.

6A-3 Deviation from ideal gas behavior occurs when attractive forces between gas molecules or molecular volumes become significant. This occurs at low temperatures or high pressures.

6A-5

P(atm)	V(L)	PV(atm·L)
1	1	1
45.8	0.01705	0.7809
84.2	0.00474	0.3991
110.5	0.00411	0.4542
176.0	0.00365	0.6424
282.2	0.00333	0.9397
398.7	0.00313	1.248

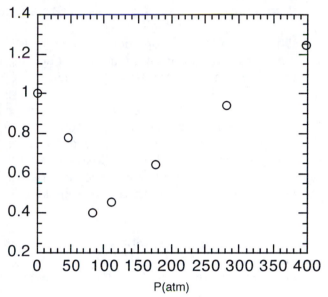

If Boyle's law was followed, PV for acetylene should be constant at a fixed temperature. Clearly, the data presented here show that this is not the case. As the pressure increases, the attractive forces between particles predominates in causing deviation from ideal behavior. As the pressure increases even more, the space occupied by individual gas particles becomes more significant. At the higher pressures, the relative particle size becomes the predominant cause of deviation.

6A-7 The van der Waals constant, b, for the gases are

He 0.02370 L/mol
Ne 0.01709 L/mol
Ar 0.03219 L/mol

The values of b are proportional to the volumes of the atoms. Using these values, we can compare the relative volumes of atoms.

$$\frac{He}{Ne} = 1.1 \text{ and } \frac{Ar}{Ne} = 1.2$$

6A-9 For CO_2 at 100 atm the negative deviation in the value of PV from the ideal gas law indicates that the $\frac{an^2}{V^2}$ term is important at this pressure. However for H_2 the positive deviation from the ideal gas law indicates that it is the size of the molecules which is important so the nb term is more important.

Chapter 7
Making and Breaking of Bonds

7-1 **Kinetic energy** is energy of motion. Gas molecules in motion are an example of kinetic energy. **Potential energy** is energy that is stored within a system. The energy stored within a chemical bond is an example of potential energy.

7-3 Energy can be transferred from one object to another by mechanical (contact) means. Rotation of a car's tires with respect to a road surface gives an automobile kinetic energy.

7-5 When chemical reactions occur, some chemical bonds may be formed. Energy is released during the formation of a chemical bond because the energy of the free atoms is greater than the energy of the bonded atoms.

7-7 The ball at the top of the building has a large amount of potential energy. Once the ball is released that potential energy is converted into kinetic energy as the ball moves closer to the ground. When the ball is brought to rest on the pavement, the kinetic energy of the falling object cannot be just "lost." It is transferred into the motions of the individual atoms and molecules in the ball and the pavement with which it is in contact. By transferring the energy to these motions, the average kinetic energy of these materials is increased, hence a rise in temperature.

7-9 The iron atoms in the warm bar would slow down, and the atoms in the cooler bar would speed up until they had the same average kinetic energy. When the average kinetic energies are the same, they have the same temperature.

7-11 As a balloon filled with helium becomes warmer, the particles within become more agitated. This induces a greater pressure on the walls, causing the balloon to expand.

7-13 **Specific heat** is the amount of energy required to raise the temperature of one gram of a substance by one degree Celsius. The **molar heat capacity** is the amount of energy required to raise the temperature of one mole of a substance by one degree Celsius.

7-15 If water and mercury had the same heat capacity, then the final temperature would be halfway between 50°C and 20°C, or 35°C. However, water has a much higher heat capacity than mercury. Thus the energy required to raise the one mole of water by one degree is greater than the energy released by the mercury as it goes down by one degree. So the temperature of the mercury must decrease even further than one degree to increase the temperature of the water by a degree. Therefore the final temperature will be less than 35°C.

7-17 State functions of a system are often associated with intensive properties of the system. Thus temperature, pressure, volume, enthalpy, and energy of an ideal gas are state functions.

7-19 By definition a property of a system is a state function if it depends only on the state of the system, not on the path used to get to that state.
The state functions are: (a), (b), (c), (d).

7-21 The first law of thermodynamics states that energy can be transferred between a system and its surroundings in the form of either heat or work. What the system loses the surroundings must gain and vice versa. The energy of a system is conserved only if no exchange with the surroundings is possible.

7-23 The expansion of a gas pushing a piston from a cylinder is an example of a system doing work on its surroundings. The energy of the system will decrease as it is transferred to the piston. A glass of warm lemonade placed in a refrigerator is an example of a system losing heat to its surroundings. The lemonade is considered to be the system losing heat to the surrounding refrigerator.

7-25 The first law of thermodynamics states that energy is conserved and it can neither be created nor destroyed. However, it can be transferred from one system to another either in the form of heat or work.

7-27 All chemical reactions involve the breaking and reforming of some kind of bonds between the either atoms, ions or molecules.

7-29 The change in enthalpy refers to the difference in relative enthalpy between the reactants and the products. At a constant pressure this is the difference in heat from the products and reactants.

7-31

a) $1 \text{ mol } H_2O \times \dfrac{131.29 \text{ kJ}}{1 \text{ mol } H_2O \text{ reacted}} = 131.29 \text{ kJ}$

b) $2 \text{ mol } H_2O \times \dfrac{131.29 \text{ kJ}}{1 \text{ mol } H_2O \text{ reacted}} = 262.58 \text{ kJ}$

c) $0.0300 \text{ mol } H_2O \times \dfrac{131.29 \text{ kJ}}{1 \text{ mol } H_2O \text{ reacted}} = 3.93 \text{ kJ}$

d) $0.0500 \text{ mol } H_2O \times \dfrac{131.29 \text{ kJ}}{1 \text{ mol } H_2O \text{ reacted}} = 6.56 \text{ kJ}$

7-33 $\Delta H^\circ = \dfrac{-46.22 \text{ kJ}}{1.00 \text{ g Mg}} \times \dfrac{24.305 \text{ g Mg}}{1 \text{ mol Mg}} = -1.12 \times 10^3 \dfrac{\text{kJ}}{\text{mol Mg}}$

7-35 Since processes (1) and (4) are the same they will have the same reaction enthalpy. Furthermore it doesn't mater whether process (1) is taken directly or if the combination of (2) and (3) are followed. The net enthalpy change will be the same, because the net reaction result is the same. This does illustrate how enthalpy is a state function.

7-37 One must know the initial state of the system (the reactants) and the final state of the system (the products). The enthalpy change is simply the enthalpy of the final state minus the enthalpy of the initial state.

7-39 Thermodynamic data are typically reported at 25°C and 1 bar of pressure (1 bar = 0.9869 atm, and the approximation is often made that 1 bar = 1 atm, but this should not be used for very precise calculations).

7-41 The ° symbol indicates the standard state. The standard state is defined by a unique set of conditions. There is only one standard state. Therefore, the standard state enthalpy change must be unique, while the enthalpy change for the many other possible states of the system can vary widely.

7-43   In an **exothermic** process energy is given off in the course of the reaction, therefore the bond strengths of the products are greater than for the reactants. In an **endothermic** process energy is absorbed by the system to make the reaction proceed, therefore the reactants have stronger bonds than the products.

7-45   The sign of ΔH indicates which side of the reaction, products or reactants, is favored by bonding forces. If ΔH is positive, energy flows into the system. If ΔH is negative, energy flows out of the system.

7-47   (a) Endothermic. Bonds are broken, but no bonds are formed. ΔH is positive.
(b) Exothermic. Bonds are formed, but no bonds are broken. ΔH is negative.
(c) Exothermic. Bonds are formed, but no bonds are broken. ΔH is negative.

7-49   Yes. The enthalpy is dependent upon the amount of substance formed. To form one mole of $CH_4(g)$ from its atoms, 1662.09 kJ of heat is released. To form two moles, 2 x 1.662.09 = 3324.18 kJ is released.

7-51   Endothermic reactions require an input of heat energy.
(a) Breaking chemical bonds requires input of energy. Endothermic.
(b) Condensation changes in phase are exothermic.

7-53 $\Delta H° = 2849.3$ kJ/mol - 1062.5kJ/mol - 1608.531kJ/mol $=178.3\,\dfrac{kJ}{mol_{rxn}}$

The reaction is endothermic.

7-55 $\Delta H° = 962.2$ kJ/mol + 2 x 498.340 kJ/mol - 3243.3 kJ/mol = $-1284.4\,\dfrac{kJ}{mol_{rxn}}$

7-57 $\Delta H° = 2$ x 5169.38 kJ/mol + 13 x 498.340 kJ/mol
-8 x 1608.531 kJ/mol - 10 x 926.29 kJ/mol = $-5313.97\,\dfrac{kJ}{mol_{rxn}}$

7-59 $\Delta H° = 6734.3$ kJ/mol + 6 x 970.30 kJ/mol - 4 x 3241.7 kJ/mol = $-410.7\,\dfrac{kJ}{mol_{rxn}}$

7-61 $\Delta H° = 2404.3$ kJ/mol + 2 x 326.4 kJ/mol - 2 x 416.3 kJ/mol - 3076.0 kJ/mol
$= -851.5\,\dfrac{kJ}{mol_{rxn}}$

7-63 $\Delta H° = 4$ x 1152.1 kJ/mol + 11 x 498.340 kJ/mol - 2 x 2404.3 kJ/mol
- 8 x 1073.95 kJ/mol = $-3310.1\,\dfrac{kJ}{mol_{rxn}}$

7-65   For $P_2(g)$ and $P_4(g)$ the average P-P bond enthalpy is found from the enthalpies of atom combination.
For P_2: (485.0 kJ/mol)/2 = 242.5 kJ/mol P-P bonds

For P_4: (1199.65 kJ/mol)/4 = 299.91 kJ/mol P-P bonds

The average P-P bond in P_4 is stronger than that in P_2.

7-67 $\Delta H° = 1076.377$ kJ/mol $+ 435.30$ kJ/mol $- 3 \times 415.52$ kJ/mol

$- B.E.(C\!-\!O) - 485.15$ kJ/mol $= +220.03 \dfrac{kJ}{mol_{rxn}} - B.E.\ (C\!-\!O)$

Even though we don't have the bond energy for C–O single bond, the reaction will be exothermic because $\Delta H°$ is a negative number and we'll be subtracting a positive number.

7-69 $\Delta H° = 4 \times \dfrac{937.86}{2}$ kJ/mol $- 2 \times 631.62$ kJ/mol $- 498.340$ kJ/mol $= +114.14 \dfrac{kJ}{mol_{rxn}}$

The bonds of the products are weaker than the bonds in the reactants. This is not surprising since there are four bonds in the reactants while only three in the products.

7-71 $H_3C\!-\!\!-\!CH_3$ This is the longest and also the weakest bond.

$H_2C\!=\!\!=\!CH_2$

$HC\!\equiv\!\!\equiv\!CH$ This is the shortest bond with the largest bond strength.

7-73 The difference between the initial and final values of the enthalpy of a system does not depend on the path used to go from one state to the other. Therefore, the enthalpy change for a reaction is the same regardless of the sequence of reactions used to calculate ΔH.

7-75

	$\Delta H°(kJ/mol_{rxn})$
$2H_2O_2(aq) \rightarrow 2H_2(g) + 2O_2(g)$	$2(+191.17)$
$2H_2(g) + 2O_2(g) \rightarrow 2H_2O(l)$	$2(-285.83)$
$2H_2O_2(aq) \rightarrow 2H_2O(l) + O_2(g)$	-189.32

$\Delta H° = -189.32$ kJ/mol$_{rxn}$

7-77

	$\Delta H°(kJ/mol_{rxn})$
$2N_2(g) + 6O_2(g) + 2H_2(g) \rightarrow 4HNO_3(aq)$	-829.4
$4HNO_3(aq) \rightarrow 2N_2O_5(g) + 2H_2O(l)$	$+280.48$
$2H_2O(l) \rightarrow 2H_2(g) + O_2(g)$	$+571.7$
$2N_2(g) + 5O_2(g) \rightarrow 2N_2O_5(g)$	$+22.8$

$\Delta H° = +22.8$ kJ/mol$_{rxn}$

7-79 $\Delta H° = -2510.0$ kJ/mol $- (-2456.1$ kJ/mol$) = -53.9$ kJ/mol$_{rxn}$

7-81 $\Delta H° = -635.09 \dfrac{kJ}{mol} - 393.509 \dfrac{kJ}{mol} - \left(-1206.92 \dfrac{kJ}{mol}\right) = +178.32 \dfrac{kJ}{mol_{rxn}}$

This is the same value as problem 7-53.

7-83 $\Delta H° = -1279.0 \dfrac{kJ}{mol} - 5.4 \dfrac{kJ}{mol} = -1284.4 \dfrac{kJ}{mol_{rxn}}$

This is the same value as problem 7-54.

7-85 $\Delta H° = 10\left(-241.818\dfrac{kJ}{mol}\right) + 8\left(-393.509\dfrac{kJ}{mol}\right) - 2\left(-126.2\dfrac{kJ}{mol}\right) = -5313.9\dfrac{kJ}{mol_{rxn}}$

7-87 $\Delta H° = 4\left(-1277.4\dfrac{kJ}{mol}\right) - \left(-2984.0\dfrac{kJ}{mol}\right) - 6\left(-285.830\dfrac{kJ}{mol}\right) = -410.6\dfrac{kJ}{mol_{rxn}}$.

This is the same value as in problem 7-59.

7-89 For iron(III) oxide:

$\Delta H° = \left(-1675.7\dfrac{kJ}{mol}\right) - \left(-824.2\dfrac{kJ}{mol}\right) = -851.5\dfrac{kJ}{mol_{rxn}}$

For chromium(III) oxide:

$\Delta H° = \left(-1675.7\dfrac{kJ}{mol}\right) - \left(-1139.7\dfrac{kJ}{mol}\right) = -536.0\dfrac{kJ}{mol_{rxn}}$

The reaction with iron (III) oxide evolves more heat per mol aluminum consumed.

7-91 Average C-H bond enthalpy = (-1662.09 kJ/mol)/4 = -415.52 kJ/mol
Average C-Cl bond enthalpy = (-1338.84 kJ/mol)/4 = -334.71 kJ/mol
The C-Cl bond length is greater than the C-H bond length, and because the bonds are both single bonds, the C-Cl bond enthalpy should have been predicted to be smaller.

7-93 ΔH is negative. More enthalpy is released than was required to break the reactant bonds.

7-95

Compound	ΔH_{ac}(kJ/mol)	# of C-C and C-H bonds	ΔH_{ac}(kJ/mol) of C-H bonds	Average C-H Bond Enthalpy
$CH_4(g)$	-1662.09	0/4	-1662.09	-415.52
$C_2H_6(g)$	-2823.94	1/6	-2473.94	-412.32
$C_3H_8(g)$	-3992.9	2/8	-3292.9	-411.61
$C_4H_{10}(g)$	-5169.38	3/10	-4119.39	-411.94

7-97 The more negative the ΔH_{ac} the stronger the bonds.

7-99

(a) $CH_3OCH_3(g) + 3\ O_2(g) \longrightarrow 2\ CO_2(g) + 3\ H_2O(g)$

$CH_3CH_2OH(g) + 3\ O_2(g) \longrightarrow 2\ CO_2(g) + 3\ H_2O(g)$

(b) For CH_3OCH_3: $\Delta H° = 2 \times (-1608.531) + 3 \times (-926.29) - (-3171.3)$
-(3 x -498.340)= -1329.6 kJ/mol$_{rxn}$
For CH_3CH_2OH: $\Delta H° = 2 \times (-1608.531) + 3 \times (-926.29) - (-3223.53)$
-(3 x -498.340)= -1277.4 kJ/mol$_{rxn}$

(c) Dimethyl ether releases more heat.

(d) Since the products are the same for both reactions and ethyl alcohol released less heat, it must have started with the stronger bonds than the dimethyl ether.

7-101 (b) This is because bond forming produces heat.

7-103 N_2: 945.408/3 = 315.136 kJ/mol; H_2 435.30 kJ/mol; N-H: 390.59 kJ/mol

ΔH = bonds broken in reactants - bonds formed in products

ΔH = 3 x (315.136 kJ/mol) + 3x (435.30 kJ/mol) - 6 x (390.59 kJ/mol)

= -92.2 kJ/mol

exothermic

7-105 ΔH = -1291.9 kJ/mol - (-455.6 kJ/mol) - 2 x (-435.30 kJ/mol) = +34.3 kJ/mol

The amount of energy released during bond formation is not great enough to compensate for the input of energy required to atomize the reactants.

7-107   (a) Cl_2 has the shorter bond so it will be stronger.

(b) IBr has the longer bond so it will be weaker.

(c) Yes

Compound	ΔH_{ac}(kJ/mol)
I_2(g)	-151
Cl_2(g)	-243
Br_2(g)	-193
ICl(g)	-211
IBr(g)	-178

(d) I_2(g) + Cl_2(g) $\longrightarrow$ 2 ICl(g)

ΔH = 151 kJ/mol + 243 kJ/mol – 2 x 211 kJ/mol = –28 kJ/mol

I_2(g) + Br_2(g) $\longrightarrow$ 2 IBr(g)

ΔH = 151 kJ/mol + 193 kJ/mol – 2 x 178 kJ/mol = –12 kJ/mol

The first reaction is more exothermic

(e) The Cl_2 bond is harder to break than the Br_2, but the I–Cl bond is also stronger than the I-Br bond and there are two formed in each reaction.

7-109   (a) Using the bond triangle from Chapter 5, one finds that Mg is metallic, Cl_2 is covalent and $MgCl_2$ is ionic.

(b)

Compound	ΔH_{ac}(kJ/mol)
Mg(s)	-147.7
Cl_2(g)	-243.4
$MgCl_2$(s)	-1032.4

ΔH = 147.7 kJ/mol + 243.4 kJ/mol – 1032.4 kJ/mol = –641.3 kJ/mol

(c) The product molecule has an ionic bond whereas the reactant molecules have covalent and metallic bonding which are much weaker, therefore it is going to be very exothermic.

Chapter 8
Liquids and Solutions

8-1   On a molecular scale gases are characterized as randomly moving particles widely separated from one another. Solids are characterized as units positioned at relatively short distances in an ordered arrangement. Liquids are characterized in terms of an intermediate arrangement of building units. There is long range ordering for solids, intermediate range ordering for liquids, and essentially no ordering beyond the molecular boundary for gases. On the macroscopic scale, solids retain their shape. A liquid conforms to the lower contours of its container. Gases expand to fill the entire container. The higher compressibility of gases versus that of liquids and solids is another macroscopic manifestation of the differences between phases of matter.

8-3   The molecules within a solid vibrate so rapidly as the solid melts that they begin to explore areas outside the rigid confines of the ordered structure of the solid. The liquid is made up of molecules that are still connected to each other by intermolecular forces. However, the kinetic energy of the molecules in a liquid is high enough that it is not possible for those molecules to maintain an "ordered" structure. As the liquid is heated to boiling, the last of the strong intermolecular forces are broken as the molecule is moving fast enough to overcome any attractions to other molecules in the liquid. Once in the gas phase, the molecule is relatively free of intermolecular forces until it, by chance, comes in contact with another molecule through a collision.

8-5   The level of intermolecular forces that exists between molecules of a substance will determine what phase the substance is found in at room temperature. If the substance has very large intermolecular attractions it will be found in the solid phase at room temperature. If the intermolecular forces are weak, the substance will be a gas at room temperature.

8-7   **Dipole-dipole** forces arise when molecules have a partial charge separation as an inherent component of the molecular structure. This type of force is purely electrostatic. The interactions between molecules of HCl, which has a large dipole moment, are primarily dipole-dipole.
     Dipole-induced dipole forces result when the charge separation associated with a molecular dipole causes (induces) an instantaneous distortion in the electron charge distribution about a molecule that does not have a dipole by itself. This distortion in turn results in an instantaneous charge separation, which in turn is attracted to the first molecular dipole in the same way a dipole-dipole force arises. An example of this would be the solvation of non-polar CCl_4 by the very polar solvent acetone.
     Induced dipole-induced dipole forces (dispersion) result from the instantaneous distortion of a molecular electron charge distribution by interaction with another molecular electron charge distribution. Dispersion forces are responsible for keeping the nonpolar CCl_4 molecules in liquid form at room temperature.
     Hydrogen bonding is a specific dipole-dipole interaction and results when hydrogen is bonded to electronegative atoms such as N, O, or F. The electronegative element withdraws a large charge density from the proton, making it a very concentrated positive charge. This concentrated positive charge is then attracted to the electronegative atom on a second molecule. In this way the hydrogen acts as a "bridge" between the two electronegative atoms. The hydrogen atom is covalently bound to the first atom and hydrogen bonded to the second. The most ubiquitous example of hydrogen bonding is water.

8-9   A hydrogen bond will only exist when there is a hydrogen atom that is covalently bound to an electronegative element such as N, O or F.

8-11   If the molecule has a dipole moment the molecule will exhibit dipole-dipole forces. If there is a non-polar part of the molecule, there will also be dipole-induced dipole forces. Dispersion forces are always present. And if the dipole of the molecule is due to an O–H, N–H or H–F bond, there will be hydrogen bonding present.

8-13   In these organic molecules, the intermolecular forces are due primarily to dispersion interactions and one should expect carbon tetrachloride to have a higher boiling temperature because it has the most polarizable electron cloud and it is has the highest molecular mass. Response (e).

8-15   Propane is a lighter molecular weight compound, so the dispersion forces are less in propane.

8-17   The dangling, drawn out structure of n-pentane allows for a closer approach of molecules, thus increasing the dispersion interactions. The structure of isopentane is more compact and symmetric. Overall the isopentane molecule is more spherical. The distance of closest approach of isopentane molecules is greater than with n-pentane and hence the dispersion interactions are weaker.

8-19   The temperature of a substance is a measure of its average kinetic energy. If two sets of the same molecules are at the same temperature, they will have the same average kinetic energy whether they are in the gas phase or liquid phase. If the molecules have the same mass then they must have the same velocity regardless of what phase they are in.

8-21   The enthalpy of vaporization is the enthalpy change required to break all the intermolecular forces holding a molecule in the liquid phase, freeing it to become a gas. The enthalpy of fusion is the enthalpy change that accompanies the breaking of some intermolecular bonds when the rigid structure of a solid is transformed to the liquid, where intermolecular forces are still present.

8-23   Vapor pressure is a function of temperature and increases with increasing temperature.

8-25   The kinetic energy of a gas is the energy it has because of the motion of the gas particles. But not all gas particles are moving at exactly the same speed. In a gaseous system at any one moment, some of the gas molecules are moving faster than others.

8-27   A cloth soaked in water feels cool when it is placed on your forehead because the process of evaporation is endothermic. The heat energy required to cause the evaporation of the water in the cloth comes from your forehead.

8-29   The liquid and vapor phases are at equilibrium when the rate of molecules evaporating is the same as the rate of molecules condensing.

8-31   Even at temperatures below the freezing point of water some (but not many) molecules on the surface of the ice will have enough energy to break free from the attractive forces of the rest of the solid. This process is what is called sublimation. It is much the same as water that evaporates from a liquid even when the liquid is below the boiling point of water.

8-33   The force between the mercury atoms is so much stronger than the force between the mercury and the glass that the mercury forms balls to reduce interaction with glass.

8-35   Car wax has a nonpolar surface to it, while water is a very polar molecule. On a freshly waxed surface there is little adhesion between water molecules and the wax molecules. Water beads up because it is more attracted to other water molecules than the surface of the car.

8-37   The molecule with the lowest intermolecular forces will have the lower boiling point and the highest vapor pressure. In each pair one will have the lower boiling point and the other will have lowest vapor pressure. The lowest boiling point is listed.
        (a) $CH_3CH_2CH_3$ (b) CH_3CH_3 (c) CH_3CH_2OH (d) $CH_3CH_2CH_2CH_2CH_3$

8-39   A solid does not always get hotter when heat is added. If the solid is at the melting temperature, then any added heat will go into breaking intermolecular bonds, thus melting the material.

8-41   The melting point of a substance is dependent upon the intermolecular forces of the substance. Materials with low intermolecular forces will have low melting points, while materials with a large amount of intermolecular forces will require more kinetic energy (and thus a higher temperature) to break these bonds.

8-43   Denver is at a higher altitude than Miami. The atmospheric pressure is less in Denver, and water boils at a lower temperature.

8-45   The vapor pressure of water reaches 50 mm Hg at 38.1 °C.

8-47   Increasing the temperature of a liquid will increase the vapor pressure of the liquid. Response (c).

8-49   With the nozzle up, gaseous butane will rise to the top and escape.   Since the butane is under pressure inside the can, some of the butane will liquefy. This liquid will drip out of the nozzle when it is pointed down.

8-51   High pressure and low temperature.

8-53   When the horizontal dotted line intersects a solid line there is a phase change. At a pressure of lower than one atm, the freezing point will be slightly less than at one atm and the boiling point will be less than at one atm. The reduction in boiling point will be larger than the reduction in melting point.

8-55   The boiling and melting temperatures of water are much higher than expected from extrapolation of the corresponding values for the dihydrogen compounds of the other members of Group VIA. The concept of the hydrogen bond was introduced to account for the experimental observations. Hydrogen bonds are strong intermolecular bonds in water that arise because of the force of attraction between positively charged hydrogen atoms on one water molecule and the negatively charged oxygen atom on another water molecule. In order to boil water, these intermolecular forces of attraction must be broken and because these forces are stronger for H_2O than the other compounds, a higher temperature is required for water molecules to escape the liquid.

8-57   The rigid hydrogen bonded network in ice results in large hexagonal "open spaces." In the liquid phase water is still hydrogen bonded, but the interactions are not as structured and molecules will flow inside the "open spaces," decreasing the volume that the same number of molecules will take up.

8-59   It takes a large amount of energy to disrupt the hydrogen bonds in water.

8-61   Anesthetics and olive oil are relatively nonpolar.

8-63   As the length of the nonpolar "tail" increases, the polar head of the molecule becomes less significant in determining the physical properties of the compound. (e) 1-heptanol, should display the greater solubility in a nonpolar solvent such as carbon tetrachloride.

8-65   Nonpolar PCl_5 will dissolve better in a nonpolar solvent. The ionic products will dissolve better in a polar solvent. The CCl_4 will favor the reactant. A polar solvent will favor the products.

8-67   Alcohols have a hydrophilic end which is readily soluble in water and a hydrophobic "tail" that is not soluble. The level to which an alcohol is soluble in water is dependent upon the size of the hydrophobic "tail" on the alcohol. The longer the hydrophobic end of the molecule the less soluble it is in water.

8-69   Water cannot always penetrate dirt on clothes if the dirt is covered by a hydrophobic layer of grease. When soap is dissolved in water, it tends to stay on the surface of the water. The polar hydrophilic end of a soap molecule will be oriented on the surface of water, while the nonpolar hydrophobic end of the soap molecule will point away from the water. However this end of the soap molecule will also dissolve the nonpolar grease or oil particles in dirty clothes, effectively removing it from the clothing.

8-71   "Dry" cleaning is done without water. Rather than using water as the solvent, a nonpolar solvent is used to "wash" clothing. The nonpolar solvent will readily dissolve grease or oil in the clothes and the soil particles can be removed by detergents dissolved in a nonpolar solvent. The advantage to this method is that delicate materials do not need to be heated to wash them and the fibers in the clothes will not swell and distort the way they do when water is used. However, this method is not without its own difficulties. Early dry cleaning solvents were quite flammable. The current solvents, halogenated hydrocarbons, must also be handled with extreme care and disposed of in an appropriate manner because of their potential danger to the environment.

8-73   Solubility properties of a solid depend upon the interplay of lattice forces acting to hold the particles together in the solid and solvation forces acting to disperse the solute components throughout the solution. In the case of $BaCl_2(s)$, the enthalpy of solution is exothermic. In the case of $AgCl(s)$, the enthalpy of solution is endothermic.

8-75   Tap water has a limited concentration of ions in solution. The solution of NaCl has a much higher concentration of ions. The electric current needed to light the bulb must pass through the solution. The larger the number of ions (charge carriers), the larger the current which can be transported through the solution.

8-77   Using the solubility rules we find that nitrates are soluble, chlorides are generally soluble, and although not specifically mentioned, acetates (OAc^-) are also generally soluble. Sulfides are usually insoluble, but BaS is an exception. Carbonates and chromates are usually insoluble. Barium carbonate (c) is insoluble.

8-79   Using the solubility rules we find that nitrates are soluble. Lead nitrate (e) is the single salt of the group that is soluble in water.

8-81   a) $Mg(s) + 2 HCl(aq) \rightarrow MgCl_2(aq) + H_2(g)$
        $Mg(s) + 2 H^+(aq) \rightarrow Mg^{2-}(aq) + H_2(g)$
     b) $Na_2CO_3(aq) + Ca(NO_3)_2(aq) \rightarrow 2 NaNO_3(aq) + CaCO_3(s)$
        $CO_3{}^{2-}(aq) + Ca^{+2}(aq) \rightarrow CaCO_3(s)$

8-83   The NaCl bonds are ionic bonds, that is, attractions between a positively charged ion and a negatively charged ion. The bonds between water molecules are an attraction of partial negative charge for a partial positive charge (hydrogen bond). The boiling points illustrate the fact that the hydrogen bond between molecules is much weaker than the ionic bonds that hold ions together.

8-85   The compound on the right (boiling point 35 °C) can form a hydrogen bond, the one on the left cannot.

8-87   As the water is heated the temperature would increase until the boiling point is reached. At this point, the temperature remains constant until all the water has vaporized. The temperature would then begin to increase. If no heat is supplied to the beaker, the temperature would remain constant. At 100°C the vaporization is an equilibrium process.

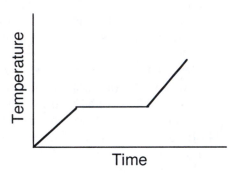

8-89 $\Delta H = \Delta E + \Delta nRT$, where ΔE is the energy of the intermolecular forces in the liquid and the ΔnRT is the work done by the system to increase the volume taken up by the gas. The second term will increase with temperature and the enthalpy change for 1 mole of gaseous water would be greater at 100°C.

8-91   (b) If a substance has a high vapor pressure, it can easily escape the liquid phase and enter the vapor phase. Since boiling occurs when the vapor pressure of a liquid equals atmospheric pressure, a substance with a high vapor pressure does not need to be heated to a high temperature for boiling to occur.

8-93   Pressure exerted by 1.00 g of benzene.

$$PV = nRT, \quad P = \frac{\left(\frac{1\,g\,C_6H_6}{78.113\,g/mol}\right) \times (0.082056\,L\,atm\,/(mol\,K)) \times (353\,K)}{(1.00\,L)} = 0.371\,atm$$

= 282 mmHg

Since the pressure of 1.00 g of gaseous benzene is less than the equilibrium vapor pressure, all of it will be in the vapor phase. Thus, 1.00 g will evaporate and the final pressure will be 282 mmHg.

In a 500 mL container,

$$PV = nRT, \quad P = \frac{\left(\frac{1\,g\,C_6H_6}{78.113\,g/mol}\right) \times (0.082056\,L\,atm\,/(mol\,K)) \times (353\,K)}{(0.500\,L)} = 0.742\,atm$$

=564 mmHg

Since the equilibrium vapor pressure is 325 mmHg at 80°C, all of the benzene will not evaporate in the 500 mL box. The liquid will be in equilibrium with the vapor.

$$mass = \frac{PV \cdot MW}{RT} = \frac{\left(\frac{325}{760}\right) \times 0.50L \times 78.113\frac{g}{mol}}{0.08206\frac{L\,atm}{mol\,K} \times 353K} = 0.577g$$

The mass of benzene in the vapor will be 0.577 grams and the final pressure will be 325 mmHg. The difference between the two experiments is related to the volume of the containers. In the smaller container, there was enough liquid benzene to achieve equilibrium.

8-95   (a) Before the valve is opened, the liquid and gas are at equilibrium, i.e. for every molecule that evaporates one condenses and the pressure is observed to remain constant. After some of the gas is released, the system is no longer at equilibrium and the pressure is lowered. After some time, however, equilibrium is re-established and the pressure remains constant.
(b) After the piston is compressed, the pressure increases because the volume has decreased at constant temperature. To re-establish equilibrium, the total gas pressure decreases to a constant value.

8-97   Vapor pressure should increase as temperature increases because the molecules in the liquid phase have greater kinetic energy and can more easily escape into the vapor. Surface tension should decrease. As the temperature is raised, it would become easier to increase the surface area of the liquid. The enthalpy of vaporization will increase with temperature based upon the amount of work done. See the answer for 8-89 for more detail.

8-99   The intermolecular forces of liquid water are greater than those in liquid methanol, therefore the water molecules experience a higher surface tension.

8-101   n-hexane   2-hexanone   2-hexanol

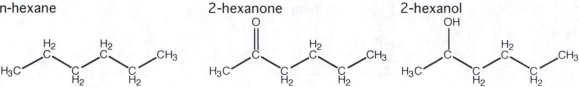

   (a) n-hexane: dispersion; 2-hexanone: dispersion, dipole-induced dipole, and dipole-dipole;
   2-hexanol: dispersion, dipole-induced dipole, dipole-dipole and hydrogen bonding
   (b) 2-hexanol because it has the greatest intermolecular forces.
   (c) 2-hexanol because it has the greatest intermolecular forces.
   (d) 2-hexanol because it has a hydrogen bonding (hydrophilic) group in the –OH.

8-103   (a) Boiling point (C) < (A) < (B) C has some polar groups in its structure, but they are
        flanked by non-polar groups making dipole-dipole forces weaker. (A) has a polar group on
        the end which allows both dipole-dipole and dipole-induced dipole forces to dominate. (B)
        has hydrogen bonding, giving this the highest boiling point.
        (b) Vapor pressure (B) < (A) < (C); this is for the opposite reasons as part (a).
        (c) (B) will be soluble in water because it can form hydrogen bonds.

8-105   In the liquid phase there are still some intermolecular forces holding the liquid together so
        not all the bonds are broken when a solid melts to form a liquid.

Chapter 8 Special Topics

8A-1 **Colligative properties** are properties of solutions that will vary with the concentration of the solute particles in the solution.

8A-3 (a) hexane and heptane would be the closest to an ideal solution.

8A-5 The vapor pressure of a pure liquid depends only upon the temperature. The vapor pressure of a solution, however, is lower than that of the pure liquid by an amount proportional to the added solute.

8A-7 First we need to find the moles of each constituent of the solution:

$$moles_{C_5H_{12}} = \frac{500g}{72.150\frac{g}{mol}} = 6.930 mole$$

$$moles_{C_7H_{16}} = \frac{500.g}{100.203\frac{g}{mol}} = 4.990 mole$$

then

$$\chi_{C_5H_{12}} = \frac{6.930 mole}{6.930 mole + 4.990 mole} = 0.581$$

$$\chi_{C_7H_{16}} = \frac{4.990 mole}{6.930 mole + 4.990 mole} = 0.419$$

and $P_T = \chi_{C_5H_{12}} \times P^\circ_{C_5H_{12}} + \chi_{C_7H_{16}} \times P^\circ_{C_7H_{16}}$,

$P_T = 0.581 \times 420 mm\,Hg + 0.491 \times 36.0 = 262 mmHg$

8A-9 Boiling occurs when the equilibrium vapor pressure of a liquid equals atmospheric pressure. Since a solution has a lower vapor pressure at a given temperature than that of the pure liquid, the solution would need to be heated to a higher temperature to boil.

8A-11 $\Delta T = k_b\,m$. If P_4 is the solute, then using the data from Table 8A.1
$T - T^\circ = k_b \times m$,

$$T = T^\circ + k_b \times m = 46.23^\circ + 2.35 \frac{^\circ C}{m} \times \frac{10.00\ g}{123.896\ g/mol} \times \frac{1}{0.0250\ kg} = 53.8\ ^\circ C$$

8A-13 $\Delta T = k_b\,m$. If I_2 is the solute, then using the data from Table 8A.1
$T - T^\circ = k_b \times m$,

$$T = T^\circ + k_b \times m = 76.75^\circ + 5.03 \frac{^\circ C}{m} \times \frac{3.41g}{253.8\ g/mol} \times \frac{1}{0.0500\ kg} = 78.1\ ^\circ C$$

8A-15 A plot of the freezing point of a solution versus molality would yield a straight line with a slope equal to the negative of the freezing depression constant.

8A-17 $\Delta T = -k_f\,m$
$T - T^\circ = -k_f \times m$,

$$T = T^\circ - k_f \times m = 0^\circ - 1.853 \frac{^\circ C}{m} \times \frac{1.00g}{212.208 g/mol} \times \frac{1}{0.0456kg} = -0.191\ ^\circ C$$

8A-19 The compound that dissolves into the greatest number of ions will produce the largest freezing point depression. Answer (c), $(NH_4)_2SO_4$ is correct.

8A-21 The salt lowers the melting point of ice, and highways can remain free of ice at temperatures lower than the freezing point of pure water.

8A-23 From the previous problem, it was shown that the percent ionization can be related to the van't Hoff factor as follows: % ionization $= 200 x \dfrac{i_{observed}}{i_{theoretical}} - 100$

For a 1.33 % ionized CH_3CO_2H solution with $i_{theoretical} = 2$, $i_{observed} = 1.013$.

$$\Delta T_f = i \times k_f \times m = 1.013 \times \left(\dfrac{1.853 °C}{m} \right) \times 0.100 \ m = 0.188 \ °C$$

The solution would freeze at -0.188°C.

8A-25 $\Delta T_b = k_b \times m$, $m = \dfrac{\Delta T_b}{k_b} = \dfrac{0.50 \ °C}{0.515 \ °C/m} = 0.97 \ m$

$$\Delta T_f = k_f \times m = \left(\dfrac{1.853 °C}{m} \right) \times 0.97 \ m = 1.8 °C$$

The solution would freeze at -1.8°C

Chapter 9
Solids

9-1   **Crystalline** solids are solids where the particles in the solid are ordered in a regular and repeating pattern from one edge of the solid to the other. **Amorphous** solids are solid materials where there is little or no ordered structure. **Polycrystalline** solids contain aspects of the previous two types. In polycrystalline solids there are many small crystalline structures that are arranged in a random fashion.

9-3   Covalent compounds are typically formed by nonmetals. Metallic compounds are formed from metals. Ionic compounds are formed from electrostatic bonds between a metal cation and a nonmetal anion.

9-5   Dispersion forces.

9-7   Molecular solids are formed from intermolecular forces holding discrete covalent molecules together. These solids are typically held together by the intermolecular forces described in Chapter 8. Network covalent solids also have covalent bonds, but they are lattice covalent bonds. That is, the atoms do not form discrete covalent molecules. They are held in a solid lattice by a network of covalent bonds. In this case the intermolecular forces are strong covalent bonds. This is why network covalent solids have such high melting points.

9-9   Diamond, graphite and fullerenes. Diamond is a covalent network solid which forms a three-dimensional lattice of carbon, making this a strong material. Graphite forms two-dimensional sheets of carbon also formed by a covalent network. However, the bonding between the sheets of graphite is not strong, making graphite easy to deform. Fullerenes are carbon structures which have a surface structure like graphite, however, the ends of the sheets wrap around and connect, making three dimensional "objects" like spheres, oblong spheres and the so-called nanotubes.

9-11   A **metallic bond** is formed between several atoms that all have a very low AVEE (or electronegativity) and share all their valence electrons. Metallic bonds are formed between elements on the left-hand side of the periodic table.

9-13   In molecular, ionic and network covalent solids the electrons are localized either on atoms or in individual bonds. In metallic bonds the electrons are delocalized.

9-15   The delocalization of valence electrons throughout a metal allows us to model the forces that act to make metals malleable and ductile. The arrangement of atoms would suggest that layers could be deformed as the spherical atoms "roll by one another." The delocalized valence electron charge distribution should correspondingly stretch and distort to accommodate the stress being applied. No rupture of bonds is needed to reorient the atom arrangement as long as the metal fabric is not broken.

9-17   The ionic solids as a class are most likely to contain a compound that is a poor conductor of electricity when solid but a very good conductor when molten.

9-19   For the indicated structures, the planes stack as

simple cubic:	AAAAAA . . .	Coordination number = 6
body-centered cubic:	ABABAB . . .	Coordination number = 8
hexagonal closest packed:	ABABAB . . .	Coordination number = 12
cubic closest packed:	ABCABC . . .	Coordination number = 12

9-21   A **coordination number** is the number of nearest neighbor atoms that "bond" to an atom in a particular crystal structure.

9-23   The picture at right shows a 3d cut-away for a body-centered cubic packing of spheres. This packing form is different from simple cubic shown above by shifting each layer of atoms "down" by half an atom. The sphere marked with an A is partially hidden in the second layer. It is directly touching a sphere to its left, right, up and down. However it is in contact with two spheres in front of it and two behind it. This gives it a coordination number of 8.

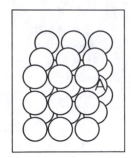

9-25   Coordination number eight is associated with the body-centered cubic structure type, (b).

9-27   A **unit cell** is the simplest repeating unit in a crystalline structure.

9-29   An intermetallic compound exists with a defined stoichiometry. An alloy represents a solid solution where the components randomly substitute one for another.

9-31   A solid solution is one where the solute atoms pack in positions normally occupied by the solvent atoms. An example of a solid solution is copper dissolved in aluminum. An interstitial solution is one type of solid solution where the solute atoms fit into the holes between the solvent atoms. Carbon dissolved in iron is an interstitial solution. To form an interstitial solution, one atom must be smaller than the other. If the atoms are the same size, the solute atoms will be too large to fit into the holes in the lattice structure, then they would be more likely to form a substitution solution.

9-33   (a), (d), (e)

9-35   **Semi-metals** are elements that have intermediate values for AVEE (or electronegativity). The AVEE is too small to effectively form covalent bonds. However, the AVEE's are usually large enough that they form only partially delocalized metallic bonds, making these elements only semi-conductive.

9-37   The lattice energy of NaCl is the energy given off in the reaction,
        (d) $NaCl(s) \rightarrow Na^+(g) + Cl^-(g)$

9-39   Larger lattice energies are associated with the lattice of a given structure type which has the positive and negative ions separated by the shortest distances. However, all else being equal, the lattice of ions of higher charge will have the larger lattice energy. MgO will have the larger lattice energy.

9-41   According to the <u>CRC Handbook of Chemistry and Physics</u>, the solubility of NaF, NaCl, NaBr and NaI in water is:
        NaF: 4.22 g/100 ml
        NaCl: 35.7 g/100 ml
        NaBr: 116.0 g/100 ml
        NaI: 184 g/100 ml
The lower solubility of NaF reflects its higher lattice energy. NaCl, NaBr, and NaI show increasing solubility in water. This corresponds to the decreasing lattice energy in these salts.

9-43 The melting points and boiling points of ionic compounds will be much higher than those of covalent compounds because the forces holding these materials together are purely ionic and strongly polar. While in covalent molecules, the intermolecular forces discussed in Chapter 8 are responsible for holding these molecules together and they are not as strong as ionic attractions.

9-45 Ceramics are formed by the fusion of silicon oxides with nonmetal atoms.

9-47 a) $\Delta EN=2.11$, average EN=2.55, Ionic, but on the border with covalent. As stated in the text, this is a ceramic material that means that it probably forms a network of bonds making it quite a hard material. However, those bonds are quite polar and this material conducts electricity.
b) $\Delta EN=0.62$, average EN=2.23, Covalent. This forms a covalent network similar to diamond (which is not surprising since Si is tetravalent like C). Also like diamond this material is very hard and doesn't conduct electricity.
c) $\Delta EN=0.49$, average EN=2.01, Semiconductor. This material will form metallic bonds, making it somewhat malleable and ductile. But because the AVEE of both elements is still somewhat high, this material will not conduct electricity very well, making it a semiconductor.
d) $\Delta EN=2.03$, average EN=2.60 , Ionic, but on the border with covalent. This material will have similar properties as CrO_2 in part a)

9-49 One needs to know the type of the unit cell.

9-51

$$\text{density}_{Ag} = \frac{4 \times \left(\dfrac{107.868 \text{ g Ag}}{1 \text{ mol}} \times \dfrac{1 \text{ mol}}{6.022 \times 10^{23} \text{ atoms}} \right)}{\left(0.40862 \text{ nm} \times \dfrac{1 \text{ m}}{10^9 \text{ nm}} \times \dfrac{100 \text{ cm}}{1 \text{ m}} \right)^3} = 10.50 \text{ g/cm}^3$$

9-53

$$\text{density}_{Ca} = \frac{N \times \left(\dfrac{40.078 \text{ g Ca}}{1 \text{ mol}} \times \dfrac{1 \text{ mol}}{6.022 \times 10^{23} \text{ atoms}} \right)}{\left(0.5582 \text{ nm} \times \dfrac{1 \text{ m}}{10^9 \text{ nm}} \times \dfrac{100 \text{ cm}}{1 \text{ m}} \right)^3} = 1.55 \text{ g/cm}^3$$

N = 4. This suggests face-centered cubic.

9-55

$$\frac{4 \times \left(\text{atomic mass} \times \dfrac{1 \text{ mol}}{6.022 \times 10^{23} \text{ atoms}} \right)}{\left(0.3608 \text{ nm} \times \dfrac{1 \text{ m}}{10^9 \text{ nm}} \times \dfrac{100 \text{ cm}}{1 \text{ m}} \right)^3} = 8.95 \text{ g/cm}^3$$

atomic mass = 62.9. The metal is copper.

9-57

$$\frac{N \times \dfrac{25.939 \text{ g LiF}}{1 \text{ mol}} \times \dfrac{1 \text{ mol}}{6.022 \times 10^{23} \text{ atoms}}}{\left(0.4017 \text{ nm} \times \dfrac{1 \text{ m}}{10^9 \text{ nm}} \times \dfrac{100 \text{ cm}}{1 \text{ m}} \right)^3} = 2.640 \text{ g/cm}^3$$

N = 4

9-59

$$V_{unit\ cell} = 2 \frac{\left(\dfrac{51.996\ g\ Cr}{1\ mol} \times \dfrac{1\ mol}{6.022 \times 10^{23}} \right)}{7.20\ \dfrac{g}{cm^3}}$$

$$V_{unit\ cell} = a^3 = 2.40 \times 10^{-23}\ cm^3$$

$$a = \sqrt[3]{2.40 \times 10^{-23}\ cm^3} = 2.88 \times 10^{-8}\ cm$$

$$d_{body}^2 = \left(\sqrt{2}a\right)^2 + a^2 = 3\,a^2$$

$$d_{body} = \sqrt{3}a = \sqrt{3} \times (2.88 \times 10^{-8}\ cm) = 5.00 \times 10^{-8}\ cm$$

$$d_{body} = 4\ r_{Cr}$$

$$r_{Cr} = \frac{5.00 \times 10^{-8}\ cm}{4} = 1.25 \times 10^{-8}\ cm$$

9-61 From the geometry of the body-centered cubic structure, we find the shortest distance between barium atoms will occur down the body diagonal where

$$d_{body} = \sqrt{3}\ a = 4r_{Ba}$$

$$r_{Ba} = \frac{\sqrt{3}}{4} \times 0.5025\ nm \times \frac{1m}{10^9 nm} \times \frac{100cm}{1m} = 2.176 \times 10^{-8}\ cm$$

The shortest distance between barium atoms will be $2\ r_{Ba}$.

The Ba-Ba distance = 4.352×10^{-8} cm.

9-63 The unit cell for the cesium chloride structure type places the $(r_{Tl} + r_l)$ distance along the body diagonal of the cube with edge distance a. From the geometry of the cubic structure type,

$$2(r_{Tl} + r_l) = \sqrt{3}\ a$$

$$r_{Tl} + r_l = \frac{\sqrt{3}}{2} \times .4198\ nm = 0.3635\ nm$$

$$r_{Tl} = 0.3635\ nm - r_l = 0.3635 - 0.216 = 0.148\ nm$$

9-65

	$\Delta H_{fus}(kJ/mol_{rxn})$	$\Delta H\ X_2(g) \rightarrow 2X(g)\ (kJ/mol_{rxn})$
F_2	0.51	157.98
Cl_2	6.41	243.358
Br_2	10.8	192.86
I_2	15.3	151.238

The force that holds the atoms together, the covalent bond, is much stronger than the dispersion forces that hold the molecules together.

9-67

1.	(e)	6.	(c)
2.	(f) or (a)	7.	(b)
3.	(d) or (b)	8.	(e)
4.	(f) or (a)	9.	(f),(a), or (c)
5.	(c) or (b)	10.	(d) or (c)
		11.	(f)

9-69   (a) BaO and MgO are both ionic and will have the higher melting points whereas HgO is a covalent species and will have a somewhat lower melting point.
(b) MgO will have a higher melting point than BaO since the internuclear distance in MgO is smaller making the ionic forces stronger than in BaO.

9-71   The tiles need to stand up to very high temperatures and must also insulate the interior of the space shuttle. The material used is SiO_2. The bond-type triangle suggests that the upper middle region along the ionic covalent interface would provide atom combinations having these characteristics.

9-73 Ni_3Al is a strong heat-resistant intermetallic compound. The structure is face-centered cubic with aluminum atoms at the corners and nickel atoms in the faces. The bond-type triangle locates this compound in the metallic region and we expect the usual properties characteristic of this type of compound. For example, it is a good conductor of electricity.

Chapter 9 Special Topics

9A-1   A **vacancy** is a point defect in a crystal lattice characterized by a missing atom in the lattice. An **interstitial impurity** is also a point defect. It occurs when an atom is forced into a space in between lattice sites. A **dislocation** occurs when a hole or vacancy is not large and a whole row of atoms in a lattice is offset from the normal lattice structure.

9A-3   If a dislocation allows one plane of atoms to slip past another plane more easily, then this would weaken the metal.

9A-5   The arrangement of atoms in a metal would suggest that layers could be deformed as the spherical atoms "roll by one another." The delocalized valence electron charge distribution should correspondingly stretch and distort to accommodate the stress being applied. No rupture of bonds is needed to reorient the atom arrangement as long as the metal fabric is not broken.

9A-7   There are a large number of closely spaced atomic orbitals associated with all of the metal atoms in a piece of metal. These orbitals have very similar energies and are often described in terms of a conduction band. An applied voltage induces electrons of the conduction band to migrate, and thereby carry an electric current.

9A-9   Arsenic atoms have one more valence electron than does silicon. If a small amount of arsenic is added to silicon, the extra electrons from the arsenic atoms occupy orbitals in a very narrow band of energies that lie between the filled and empty bands of the semiconductor. This decreases the amount of energy required to excite an electron. Gallium atoms have one less electron than does silicon. If a small amount of gallium is added to silicon, the presence of "holes" in the filled band allows the solid to carry an electric current.

9A-11   Metals are good conductors of heat. Even when they are at room temperature, the touch of your warm hand to the metal provides the metal will some heat from your hand. However, that heat is quickly drawn away because of the high conductivity.

9A-13   Both materials are covalently bound. However, the concrete is a covalent network solid, therefore the material has a greater ability to conduct heat.

9A-15   A bimetallic strip bends toward the side with the lower thermal expansion. In the case of iron and copper, iron has a smaller thermal expansion, so the strip would bend toward the iron.

9A-17   Ceramics are typically found on the border between the covalent part of the bond-type triangle and the ionic part of the triangle.

9A-19   b, j, l all might be ceramics. You are looking for covalent compounds that have ionic bonding as well.

9A-21   Ceramics that are not brittle and can take high heat have been developed. Engines made from ceramic powders operate at very high temperatures and are therefore more efficient than steel engines. These ceramic engines do not require radiators and are also much lighter than steel engines.

Chapter 10
An Introduction to Kinetics and Equilibrium

10-1 Reactions that go to completion leave no unreacted limiting reagent behind. Reactions that go to equilibrium may leave an appreciable amount of unreacted starting materials.

10-3 The **equilibrium constant expression** is the relationship between the reactant and product concentrations at equilibrium. It generally has the form of "products over reactants". The **equilibrium constant** is the numerical value obtained when all the values for the equilibrium concentrations are inserted into the equilibrium constant expression.

10-5 For the reaction A → B, the equilibrium constant expression will be $K_c = \dfrac{[B]}{[A]}$. If K_c is greater than one then the concentration of B must be greater than that of A. Conversely if K_c is less than one then the concentration of A must be greater than that of B.

10-7 The rate of a chemical reaction is equivalent to the measured change in amount of a reactant or product divided by the measured time interval over which the change in amount occurred.

10-9 The **rate constant** is the proportionality constant between the rate of the reaction and the concentrations of reactants in the rate law. The rate constant is only one part of the entire rate law.

10-11 Since the rate of the forward reaction is dependent upon the concentrations of the reactants, as those reactants are used up in the reaction, the forward reaction slows down. Conversely as the reaction proceeds there are more products present, and the rate of the reverse reaction will be dependent upon those concentrations. Therefore the rate of the reverse reaction will increase as the reaction proceeds forward.

10-13 Equilibrium was initially defined as when the forward reaction appears to stop and the concentrations of the products and the reactants do not change. However a more correct definition of equilibrium is when the rate of the forward reaction is equal to the rate of the reverse reaction.

10-15 (d) $K_c = \dfrac{[ClF_3]^2}{[Cl_2]\,[F_2]^3}$

10-17 (a) $K_c = \dfrac{[OF_2]^2}{[O_2]\,[F_2]^2}$ 　(b) $K_c = \dfrac{[SO_3]^2}{[O_2]\,[SO_2]^2}$ 　(c) $K_c = \dfrac{[SO_2Cl_2]^2[O_2]}{[SO_3]^2\,[Cl_2]^2}$

10-19 For reaction (a)

$$K_c = \frac{[NO]^2[O_2]}{[NO_2]^2}$$

For reaction (b)

$$K_c = \frac{[NO_2]^2}{[NO]^2[O_2]} = 6.2 \times 10^5$$

The value of K_c for reaction (a) is $= \dfrac{1}{K_c(\text{for reaction (b)})} = \dfrac{1}{6.2 \times 10^5} = 1.6 \times 10^{-6}$

10-21 $K_c(c) = K_c(a) \times K_c(b) = \dfrac{[CO_2]}{[CO][O_2]^{\frac{1}{2}}} \times \dfrac{[H_2][O_2]^{\frac{1}{2}}}{[H_2O]} = \dfrac{[CO_2][H_2]}{[CO][H_2O]}$

$= (1.1 \times 10^{11}) \times (7.1 \times 10^{-12}) = 0.78$

10-23 $K_c = \dfrac{[HI]^2}{[H_2][I_2]}$

trial I $\quad K_c = \dfrac{(0.015869)^2}{(0.0032583)(0.0012949)} = 59.686$

trial II $K_c = \dfrac{(0.013997)^2}{(0.0046981)(0.0007014)} = 59.45$

trial III $K_c = \dfrac{(0.005468)^2}{(0.0007106)(0.0007106)} = 59.21$

K_c is a constant.

10-25 When $Q_c > K_c$ The reaction must shift to the left to reach equilibrium. Answer (c).

10-27 $K_c = \dfrac{[COCl_2]}{[CO][Cl_2]} = 1.5 \times 10^4 \qquad Q_c = \dfrac{(0.0040)}{(0.00021)(0.00040)} = 4.8 \times 10^4$

$Q_c > K_c$ The system is not at equilibrium and must shift to the left to form reactant.

10-29 One change in N_2 concentration corresponds to 3 changes in H_2 concentration.

$$\frac{\Delta(H_2)}{\Delta(N_2)} = \frac{3}{1} \text{ and } 1 \, \Delta(H_2) = 3 \, \Delta(N_2)$$

10-31

$$2\,NH_3(g) \quad \rightleftharpoons \quad N_2(g) \quad + \quad 3\,H_2(g)$$
$$-2\,\Delta C \qquad\qquad +\Delta C \qquad\qquad +3\Delta C$$

$2\,\Delta C = 0.234$
$\Delta C = 0.117$
$\Delta(N_2) = 0.117$
$\Delta(H_2) = 0.351$

10-33 (e) $\Delta(F_2) = 3\ \Delta(Cl_2)$

10-35

	$N_2(g)$	+	$3\ H_2(g)$	$\rightleftarrows$	$2\ NH_3(g)$
initial	1.000 M		1.000 M		0
change	$-\Delta C$		$-3\Delta C$		$2\Delta C$
equilibrium	0.922 M		1.000-3(0.078)		2(0.078)

$[N_2]_{eq} = 0.922$ M $[H_2]_{eq} = 0.766$ M $[NH_3]_{eq} = 0.156$ M

10-37

	$N_2O_4(g)$	$\rightleftarrows$	$2\ NO_2(g)$	$K_c = 5.8 \times 10^{-5}$
initial	0.100 M		0	
change	ΔC		$2\Delta C$	
equilibrium	0.100 - ΔC		$2\Delta C$	

$$\frac{(2\Delta C)^2}{(0.100 - \Delta C)} = 5.8 \times 10^{-5}$$

Assume $\Delta C \ll 0.100$
$4\ \Delta C^2 = 5.8 \times 10^{-6}$
$\Delta C = 1.2 \times 10^{-3}$

Check, $\dfrac{1.20 \times 10^{-3}}{0.100} \times 100\% = 1.2\%$, the assumption is valid.

$[N_2O_4]_{eq} = 0.099$ M $[NO_2]_{eq} = 2.4 \times 10^{-3}$ M

Check, $\dfrac{(2.4 \times 10^{-3})^2}{0.099} = 5.8 \times 10^{-5}$, which agrees with the given value.

10-39

	$N_2(g)$	+ 3 $H_2(g)$	$\rightleftarrows$	$2\ NH_3(g)$	$K_c = 0.040$ at
					500°C
initial	0.10	0.10		0.10	

$Q_c = 1.00 > K_c$, reaction proceeds to the left:

change	$+\Delta C$	$+3\Delta C$		$-2\Delta C$
equilibrium	0.10 $+\Delta C$	0.10$+3\Delta C$		0.10 $-2\Delta C$

$$K_c = \frac{(0.10 - 2\Delta C)^2}{(0.10 + \Delta C)(0.10 + 3\Delta C)^3} = 0.040$$

ΔC is not small compared to 0.10. So we solve the problem by successive approximation. We find $\Delta C \approx 0.045$ and

$[NH_3] = 0.01$ M $[N_2] = 0.15$ M $[H_2] = 0.24$ M

10-41

$$N_2(g) \quad\quad + \quad O_2(g) \quad\quad\quad \rightleftarrows \quad 2NO(g)$$

0.100 - ΔC   0.090 - ΔC   2ΔC

$K_c = 3.3 \times 10^{-10}$

$$K_c = 3.3 \times 10^{-10} = \frac{(2\Delta C)^2}{(0.100 - \Delta C)(0.090 - \Delta C)}$$

Assume ΔC is very small.

$$3.3 \times 10^{-10} = \frac{4\Delta C^2}{(0.100)(0.090)}$$

$\Delta C = 8.6 \times 10^{-7}$

The assumption was valid.

$[N_2] = 0.10\ M$ $[O_2] = 0.090\ M$ $[NO] = 1.7 \times 10^{-6}$

Check: $\dfrac{(1.7 \times 10^{-6})^2}{(0.100)(0.090)} = 3.3 \times 10^{-10}$

10-43

	2NO(g)	+	Cl$_2$(g)	⇌	2 NOCl(g)
initial	0.50 M		0.10 M		0
change	-2ΔC		-ΔC		2ΔC
equilibrium	0.50 -2ΔC		0.10 - ΔC		2ΔC

$K_c = 2.1 \times 10^3$ at 500 K

The equilibrium constant is large so that the reaction will go nearly to completion. Cl$_2$ is the limiting reagent. Therefore, 0.20M of NOCl can be formed leaving 0.30M NO at equilibrium. The actual [Cl$_2$] is about 2×10^{-4} M at equilibrium.

10-45

$$\text{PCl}_5(g) \rightleftarrows \text{PCl}_3(g) + \text{Cl}_2(g) \qquad K_c = 0.0013 \text{ at } 450K$$

	PCl$_5$(g)	PCl$_3$(g)	Cl$_2$(g)
initial	1.00 M	0	0
change	$-\Delta C$	ΔC	ΔC
equilibrium	$1.00 - \Delta C$	ΔC	$0.20 + \Delta C$

$Q_c = 0 < K_c$. So the reaction will proceed to the right.

$$\frac{(\Delta C)(0.20 + \Delta C)}{(1.00 - \Delta C)} = 0.0013$$

Assume $\Delta C << 0.20$

$0.20\,\Delta C = 0.0013$

$\Delta C = 6.5 \times 10^{-3}$

Check, $\dfrac{6.5 \times 10^{-3}}{0.20} \times 100\% = 3.3\%$, the assumption is valid.

$[\text{PCl}_5]_{eq} = 0.99$ M

$[\text{PCl}_3]_{eq} = 6.5 \times 10^{-3}$ M

$[\text{Cl}_2]_{eq} = 0.21$ M

% decomposition PCl$_5$ = $\dfrac{6.5 \times 10^{-3}}{1.00} \times 100\% = 0.65\%$

The results of % decomposition calculated in section 10.7 refers to a different temperature.

10-47

$$2\,\text{NO}_2(g) \rightleftarrows 2\,\text{NO}(g) + \text{O}_2(g) \qquad K_c = 3.4 \times 10^{-7} \text{ at } 200K$$

	NO$_2$(g)	NO(g)	O$_2$(g)
initial	0.10 M	0	0.050 M
chang	$-2\Delta C$	$2\Delta C$	ΔC
equil	$0.10 - 2\Delta C$	$2\Delta C$	$0.05 + \Delta C$

$$\frac{(2\Delta C)^2 (0.050 + \Delta C)}{(0.10 - 2\Delta C)^2} = 3.4 \times 10^{-7}$$

Assume $\Delta C << 0.050$

$$\frac{(4\Delta C^2)(0.05)}{(0.10)^2} = 3.4 \times 10^{-7}$$

$\Delta C = 1.3 \times 10^{-4}$

Check, $\dfrac{2(1.3 \times 10^{-4})}{0.10} \times 100\% = 0.26\%$, the assumption is valid.

$[\text{NO}_2]_{eq} = 0.10$ M $\quad [\text{NO}]_{eq} = 2.6 \times 10^{-4}$ M $\quad [\text{O}_2]_{eq} = 0.050$ M

Check, $\dfrac{(2.6 \times 10^{-4})^2 (0.050)}{(0.099)^2} = 3.4 \times 10^{-7}$ The calculation agrees with the value of K_c given in the problem.

10-49 $2\ SO_2(g) + O_2(g) \rightleftarrows 2\ SO_3(g)\ \ K_c = 6.3 \times 10^9$

The equilibrium constant for this reaction is so large that the reaction will essentially go to completion. Thus, the concentration of SO_3 will be determined by how much SO_2 and O_2 are initially present. These two reagents are present in stoichiometric amounts so initially $SO_3 = \dfrac{0.100\ mol}{0.250\ L} = 0.400\ M$ and $O_2 = \dfrac{0.050\ mol}{0.250\ L} = 0.20\ M$, thus, a concentration of $[SO_3]$ of 0.40 M can be made.

10-51   By making a reasonable assumption that ΔC is very small compared to the initial values, mathematical equations and their solutions are simplified.

10-53   (a) The forward reaction does not proceed very far because the equilibrium constant is very small, indicating that the denominator (concentrations of reactions) in the equilibrium constant expression is large. Therefore the concentration of the product (ΔC) will be small.

(b) The equilibrium constant for the reverse reaction will be $\dfrac{1}{K_f} = 3.0 \times 10^9$. With such a large equilibrium constant, ΔC will be very large.

10-55   As the reaction system approaches more closely equilibrium state, the concentrations of all components will tend to change less and less as Q_c gets closer to K_c.

10-57   When the reaction quotient is very different from the value of the equilibrium constant, force the reaction to completion in the direction favored by the equilibrium constant and then let the reaction come back to equilibrium.

10-59   Equilibrium constants will change with temperature depending upon the nature of the reaction involved.

10-61   If decreasing the temperature decreases the equilibrium constant, then increasing the temperature should cause an increase in the equilibrium constant, which means the products will be more favored.

10-63   When the pressure changes on a reaction with gaseous components which was initially in a state of equilibrium, the reaction system shifts in the direction which will best offset the externally applied pressure change. A decrease in pressure will favor the shift in reaction direction that generates more gaseous components. An increase in pressure will cause the reaction to shift in the direction that has the smaller amounts of gaseous components. With an increase in pressure for the reactions listed:

(a) The reaction will shift to the right in the direction of increased production of SO_2Cl_2 and O_2.
(b) The reaction will shift to the right in the direction of forming OF_2.
(c) The reaction will shift to the right in the reaction of formation of NO_2.

10-65   When the concentration of a component of a reaction in equilibrium is changed the reaction system will shift in the direction to reduce the effect of the change in concentration. With an increase in the concentration of the bold component, the reaction will shift in the direction that will decrease the concentration of that component (and all other components on the same side of the reaction equation) while increasing the concentrations of all components on the other side of the reaction equation.
(a) Increasing the concentration of $NO_2(g)$ will shift the reaction to the right until equilibrium is reached.
(b) Increasing the concentration of $O_2(g)$ will shift the reaction to the left until equilibrium is reached.
(c) Increasing the concentration of $PF_5(g)$ will shift the reaction to the right until equilibrium is reached.

10-67   To increase the yield of ammonia in the Haber process, increase the pressure, increase the amount of N_2 and/or H_2 in the reaction vessel, remove NH_3 as soon as it forms.

10-69   An increase in the volume of the container by a factor of 2 would decrease the concentrations of all components of the reaction and also the pressure of the system.
$$4\ NH_3(g) + 5\ O_2(g) \rightleftarrows\ 4\ NO(g) + 6\ H_2O(g)$$
The decreased pressure would shift the equilibrium to the right in the direction of the greater number of moles of gaseous products.

10-71 $Ag_2CrO_4\ (s) \rightleftarrows\ 2\ Ag^+(aq) + CrO_4^{2-}(aq)$
$[Ag^+] = 2\ [CrO_4^{2-}]$ $K_{sp} = [Ag^+]^2[CrO_4^{2-}]$

10-73   The concentration of the pure solid, [AgCl], represents the number of moles of AgCl in a liter of solid AgCl and this is a constant. The concentrations of the ions, $[Ag^+]$ and $[Cl^-]$, represent the concentrations of the ions in moles of ion per liter of solution. These concentrations are variable within rather wide ranges because, for one thing, the ions exist in the solution phase and the volume of the solution can vary widely.

10-75   The solubility product expression for aluminum sulfate in water is
$K_{sp} = [Al^{3+}]^2[SO_4^{2-}]^3$. Response (c).

10-77 $Ag_2CO_3\ (s) \rightleftarrows\ 2\ Ag^+(aq) + CO_3^{2-}(aq)$
$[Ag^+] = 2\ [CO_3^{2-}]$ $K_{sp} = [Ag^+]^2[CO_3^{2-}]$

10-79 $SrF_2\ (s) \rightleftarrows\ Sr^{2+}(aq) + 2\ F^-(aq)$
$K_{sp} = [Sr^{2+}]\ [F^-]^2$
$K_{sp} = (8.52 \times 10^{-4})(2 \times 8.52 \times 10^{-4})^2 = 2.47 \times 10^{-9}$

10-81   (a) $Mg(OH)_2\ (s) \rightleftarrows\ Mg^{2+}(aq) + 2\ OH^-(aq)$
(b) $K_{sp} = [Mg^{2+}]\ [OH^-]^2$
(c) $K_{sp} = s\ (2\ s)^2 = 1.8 \times 10^{-11}$; s= 1.7×10^{-4} mol/L
(d) Based upon the solubility above, 1.7×10^{-5} mol will dissolve in 100 mL. $Mg(OH)_2$ has a MW of 58.318 g/mol, so we expect 9.6×10^{-4} grams to dissolve in 100 mL.

10-83 (a) $K_{sp} = [Cu^+]^2 [S^{2-}] = 4 C_s^3 = 2.5 \times 10^{-48}$ $C_s = 8.5 \times 10^{-17}$ mol/L

$$\frac{8.5\times10^{-17} \text{ mol } Cu_2S}{1000 \text{ ml soln}} \times \frac{159.158 \text{ g } Cu_2S}{1 \text{ mol } Cu_2S} = \frac{1.4\times10^{-14} \text{ g } Cu_2S}{1000 \text{ ml soln}} = \frac{1.4\times10^{-15} \text{ g } Cu_2S}{100 \text{ ml soln}}$$

(b) $K_{sp} = [Cu^{2+}][S^{2-}] = C_s^2 = 6.3 \times 10^{-36}$ $C_s = 2.5 \times 10^{-18}$ mol/L $= [Cu^{2+}]=[S^{2-}]$

$$\frac{2.5\times10^{-18} \text{ mol } CuS}{1000 \text{ ml soln}} \times \frac{95.612 \text{ g } CuS}{1 \text{ mol } CuS} = \frac{2.4\times10^{-16} \text{ g } CuS}{1000 \text{ ml soln}} = \frac{2.4 \times 10^{-17} \text{ g } CuS}{100 \text{ ml soln}}$$

10-85 $K_{sp} = [Ag^+][Br^-] = 5.0 \times 10^{-13}$

In this system the amount of Br^- already in solution will be much greater than the amount that will be added upon dissolving the AgBr so $[Ag^+]=C_s$

$K_{sp} = C_s \times 0.050 = 5.0 \times 10^{-13}$

$C_s = 1.0 \times 10^{-11}$ mol/L

$$\frac{1.0\times10^{-11} \text{ mol AgBr}}{1000 \text{ ml soln}} \times \frac{187.77 \text{ g AgBr}}{1 \text{ mol AgBr}} = \frac{1.9\times10^{-9} \text{ g AgBr}}{1000 \text{ ml soln}}$$

10-87 If $[Mg^{2+}] = \Delta C$ then $[F^-] = 2 \Delta C$ $K_{sp} = [Mg^{2+}] [F^-]^2 = [\Delta C] [2 \Delta C]^2 = 4 \Delta C^3$

Answer (e).

10-89 Let $\Delta C = [S^{2-}]$

For Ag_2S $6.3 \times 10^{-50} = 4 \Delta C^3$ $\Delta C = 2.5 \times 10^{-17} \dfrac{mol}{L}$

For HgS $4 \times 10^{-53} = \Delta C^2$ $\Delta C = 6 \times 10^{-27} \dfrac{mol}{L}$

Silver sulfide is more soluble.

10-91 Let $C_s = [S^{2-}]$

For Hg_2S $1.0 \times 10^{-47} = C_s^2$ $C_s = 3.2 \times 10^{-24} \dfrac{mol}{L}$

For HgS $4 \times 10^{-53} = C_s^2$ $C_s = 6 \times 10^{-27} \dfrac{mol}{L}$

Mercury(I) sulfide is more soluble.

10-93 $K_{sp} = 1.1 \times 10^{-12} = [Ag^+]^2[CrO_4^{2-}]$

$= [1.3 \times 10^{-4}]^2[CrO_4^{2-}]$

$[CrO_4^{2-}] = 6.5 \times 10^{-5} \dfrac{mol}{L}$

10-95 $K_{sp} = [Ag^+] [CH_3CO_2^-] = \Delta C^2$

$$\frac{1.190 \text{ g } AgCH_3CO_2}{99.40 \text{ ml}} \times \frac{1000 \text{ ml}}{1 \text{ L}} \times \frac{1 \text{ mol } AgCH_3CO_2}{166.91 \text{ g } AgCH_3CO_2} = \Delta C$$

$\Delta C = 7.173 \times 10^{-2} \dfrac{mol}{L}$

$K_{sp} = 5.145 \times 10^{-3}$

10-97 $K_{sp} = [Ba^{2+}] [SO_4^{2-}] = \Delta C^2$

Taking the density to be 1g/ml gives : $400{,}000 \text{ g} \times \dfrac{1 \text{ ml}}{1 \text{ g}} = 400{,}000 \text{ ml} = 400 \text{ L}$

$$\dfrac{1 \text{ g BaSO}_4}{400 \text{ L soln}} \times \dfrac{1 \text{ mol BaSO}_4}{233.39 \text{ g BaSO}_4} = \dfrac{1 \times 10^{-5} \text{ mol BaSO}_4}{1 \text{ L soln}} = C_s$$

$K_{sp} = 1 \times 10^{-10}$

10-99 (a) $K_{sp} = \Delta C^2 = 1.0 \times 10^{-47}$ $\Delta C = 3.2 \times 10^{-24}$ mol/L

$$\dfrac{3.2 \times 10^{-24} \text{ mol Hg}_2\text{S}}{1000 \text{ ml soln}} \times \dfrac{433.25 \text{ g Hg}_2\text{S}}{1 \text{ mol Hg}_2\text{S}} = \dfrac{1.4 \times 10^{-21} \text{ g Hg}_2\text{S}}{1000 \text{ ml soln}} = \dfrac{1.4 \times 10^{-22} \text{ g Hg}_2\text{S}}{100 \text{ ml soln}}$$

(b) $K_{sp} = \Delta C^2 = 4 \times 10^{-53}$ $\Delta C = 6 \times 10^{-27}$ mol/L

$$\dfrac{6 \times 10^{-27} \text{ mol HgS}}{1000 \text{ ml soln}} \times \dfrac{232.66 \text{ g HgS}}{1 \text{ mol HgS}} = \dfrac{1 \times 10^{-24} \text{ g HgS}}{1000 \text{ ml soln}} = \dfrac{1 \times 10^{-25} \text{ g HgS}}{100 \text{ ml soln}}$$

10-101 Let ΔC represent the solubility of the salt.

(a)	Ag$_2$S	$K_{sp} = 4\Delta C^3$	$\Delta C = 2.5 \times 10^{-17}$ mol/L
(b)	Bi$_2$S$_3$	$K_{sp} = 108\Delta C^5$	$\Delta C = 2 \times 10^{-20}$ mol/L
(c)	CuS	$K_{sp} = \Delta C^2$	$\Delta C = 2.5 \times 10^{-18}$ mol/L
(d)	HgS	$K_{sp} = \Delta C^2$	$\Delta C = 6 \times 10^{-27}$ mol/L

In order of increasing solubility: HgS < Bi$_2$S$_3$ < CuS < Ag$_2$S

10-103 $Q_{sp} = [Pb^{2+}] [F^-]^2 = 6.9 \times 10^{-12}$

$K_{sp} > Q_{sp}$, therefore a precipitate will not form.

10-105 Based upon dilution calculations, the final concentrations of Ag$^+$ and CO$_3^{2-}$ will both be 0.0010 M.

$Q_{sp} = [Ag^+]^2 [CO_3^{2-}] = 1.0 \times 10^{-9}$

$K_{sp} < Q_{sp}$, therefore a precipitate will form.

10-107 The addition of sodium chloride to a saturated solution of silver chloride will have the effect of placing stress on the solubility equilibrium. The added chloride ion will force the precipitation of silver chloride to decrease the overall concentration of chloride ions in solution. Both the chloride and silver ion concentrations will decrease from those following addition of NaCl. However in terms of the prior equilibrium concentrations, the silver ion concentration will decrease while the chloride ion concentration will exceed the prior equilibrium concentrations because extra chloride ions have been added.

10-109 $K_{sp} = [Ag^+] [I^-] = 8.3 \times 10^{-17}$

$$\dfrac{3.21 \text{ g KI}}{0.350 \text{ L}} \times \dfrac{1 \text{ mol KI}}{166.00 \text{ g KI}} = 5.53 \times 10^{-2} \dfrac{\text{mol}}{\text{L}} = [I^-]$$

$$[Ag^+] = \dfrac{8.3 \times 10^{-17}}{0.0553} = 1.5 \times 10^{-15} \dfrac{\text{mol}}{\text{L}}$$

10-111 (a) $K_{sp} = [Pb^{2+}][OH^-]^2 = 1.2 \times 10^{-15}$

$$\frac{1.2 \times 10^{-15}}{(1.0 \times 10^{-7})^2} = [Pb^{2+}] = 0.12 \text{ mol/L}$$

(b) $\dfrac{1.2 \times 10^{-15}}{(0.010 \text{ mol/L})^2} = [Pb^{2+}] = 1.2 \times 10^{-11} \text{ mol/L}$

(a) has the greater solubility

10-113 The correct response will show some of the MX_2 dissolving and producing M^{2+} and X^- in a 1 to 2 stoichiometric ratio. Response (c) matches these criteria.

10-115 $K_c = \dfrac{k_A}{k_B} = \dfrac{[B]}{[A]}$

(a) If $k_A = k_B$, then $[B] = [A]$ and $K_c = 1$, graph (3).
(b) If $k_A = 2 \times k_B$, then $[B] = 2 \times [A]$ and $K_c = 2$, graph (2).
(c) If $k_A = 1/2 \times k_B$, then $[A] = 2 \times [B]$ and $K_c = 1/2$, graph (4).

10-117 The equilibrium constant determined from the graph (a) would yield a value less than 1. To differentiate between the final two graphs, calculate Q from the stated initial concentrations and compare to the given K. Data from graph (c) yields Q < K. Thus, more B will form and A will be consumed. Q for graph (b) is equal to K; therefore, the reaction would be at equilibrium and no change in concentrations would be expected. Graph (b) is the correct representation.

10-119 a) $K_{sp} = [Ca^{2+}][SO_4^{2-}] = (4.9 \times 10^{-3})^2 = 2.4 \times 10^{-5}$
b) $Q = [Ca^{2+}][SO_4^{2-}] = (0.0036)(0.0080) = 2.88 \times 10^{-4} > K_{sp}$.
Since $Q > K_{sp}$, the reaction shifts to the left and $CaSO_4$ will precipitate.
c) From part a) the equilibrium concentrations of Ca^{2+} and SO_4^{2-} will both be 4.9×10^{-3}. This is a concentration so it doesn't matter whether there is one liter or 500 mL or even 10 mL. The concentration at equilibrium will be the same in both cases.

10-121 Graphs (b) and (e) are possible. The rate constant for the forward reaction is greater than that of the reverse reaction. This means that at equilibrium there should be more product (trans) than reactant (cis). Both graphs (b) and (e) depict an equilibrium region where there is more trans than cis isomer.

10-123 (a) A reaction is said to be at equilibrium when the rate of the forward reaction equals the rate of the reverse reaction. As a result, the concentration of reactants and products appears to remain constant. All reactions eventually come to equilibrium. It's just that some reactions have very slow reverse reactions.

(b) $K_c = \dfrac{[NO_2]^2}{[NO]^2 [O_2]} = \dfrac{(15.5)^2}{(0.0542)^2 (0.127)} = 6.44 \times 10^5$

(c) $Q = \dfrac{[NO_2]^2}{[NO]^2 [O_2]} = \dfrac{(16.5)^2}{(0.0542)^2 (1.127)} = 8.22 \times 10^4$

Since Q<K, the reaction will proceed to the right.

10-125 In each figure, there is an even number of Cl atoms represented. These would have been produced from the dissociation of Cl_2. Since no temperature is specified, the value of K cannot be specified. Therefore, each of the figures is a valid representation of the system.
Answer (e).

Chapter 11
Acids and Bases

11-1 An **acid** is a proton donor. A **base** is a proton acceptor.
Vinegar and lemon juice are examples of acids.
Lye is an example of a base.

11-3 Litmus will turn red in the presence of an acid. Litmus will remain blue in the presence of a base.

11-5 (a) and (b) are Arrhenius acids.

11-7 $HBr(g) + H_2O(l) \rightarrow H_3O^+(aq) + Br^-(aq)$

$H_2O(l) + NH_3(g) \rightarrow OH^-(aq) + NH_4^+(aq)$

11-9

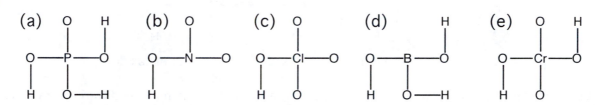

11-11 A **Brønsted base** is a substance that can accept a proton, i.e., a H^+ ion. The hydronium ion H_3O^+ with the positive charge will not add a second proton. H_3O^+ and BH_4^- are the only ions listed which cannot be Brønsted bases.

11-13 (a) $HSO_4^-(aq) +$ $H_2O(l)$ $\rightarrow$ $H_3O^+(aq) +$ $SO_4^{2-}(aq)$
      Brønsted acid   Brønsted base   Brønsted acid Brønsted base
(b) $CH_3CO_2H(aq) + OH^-(aq)$ $\rightarrow$ $CH_3CO_2^-(aq) + H_2O(l)$
      Brønsted acid   Brønsted base   Brønsted base Brønsted acid
(c) $CaF_2(s) +$ $H_2SO_4(aq)$ $\rightarrow$ $CaSO_4(aq) +$ $2 HF(aq)$
      Brønsted base Brønsted acid   Brønsted base Brønsted acid
(d) $HNO_3(aq) +$ $NH_3(aq)$ $\rightarrow$ $NH_4NO_3(aq)$
      Brønsted acid   Brønsted base

The NH_4^+ ion of $NH_4NO_3(aq)$ is a Brønsted acid.

The NO_3^- ion of $NH_4NO_3(aq)$ is a Brønsted base.

(e) $LiCH_3(l) +$ $NH_3(l)$ $\rightarrow$ $CH_4(g) +$ $LiNH_2(s)$

CH_3^- from $LiCH_3(l)$ is a Brønsted base.

$NH_3(l)$ is a Brønsted acid in this reaction.

CH_4 is an acid and $LiNH_2$ is a base.

11-15

(a) HCl(aq)   +   H₂O(l)   →   H₃O⁺(aq)   +   Cl⁻(aq)
     acid   base   acid   base
(b) HCO₃⁻(aq)   H₂O(l)   →   OH⁻(aq)   +   H₂CO₃(aq)
     base   acid   base   acid
(c) NH₃(aq)   +   H₂O(l)   →   OH⁻(aq)   +   NH₄⁺(aq)
     base   acid   base   acid
(d) CaCO₃(s)   +   2 HCl(aq)   →   Ca²⁺(aq)   +   2Cl⁻(aq)   +   H₂CO₃(aq)
     base   acid   base   acid

11-17

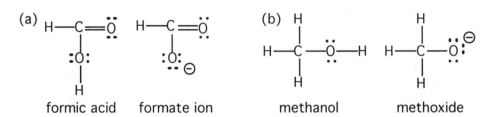

formic acid   formate ion   methanol   methoxide

11-19 The conjugate base of a Brønsted acid will have a molecular formula differing from that of the acid by the loss of one H⁺.
(a) H₂O   (b) OH ⁻   (c) O²⁻   (d) NH₃

11-21 The conjugate acid of a Brønsted base will have a molecular formula differing from that of the bases by addition of one H⁺.
(a) OH⁻   (b) H₂O   (c) H₃O⁺   (d) NH₃

11-23 This statement is true because we know that water will dissociate in the following reaction: H₂O (l) + H₂O (l) → H₃O⁺(aq) + OH⁻(aq). Since one of the water molecules is donating a proton, this is a Brønsted acid. The other water molecule in this reaction is accepting the proton, and is therefore a Brønsted base.

11-25

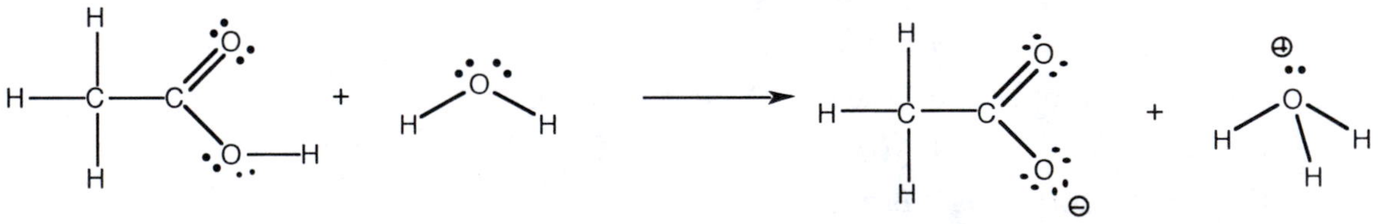

11-27 Water cannot be at equilibrium unless
       $K_W = [H_3O^+] [OH^-]$
If water contained large quantities of both ions, the reaction product, K_W, would far exceed the limiting value of 1.0×10^{-14}.

96

11-29 Addition of a strong acid or base to water increases the concentration of either the H_3O^+ ion or the OH^- ion, respectively. This shifts the dissociation reaction of water, shown below, to the left. $2H_2O$ (l) $\rightleftarrows$ H_3O^+ (aq) + OH^- (aq)

11-31 When an acid is added to water the two sources of H_3O^+ are from the reaction of the acid with water (the acid donates a proton) and from the dissociation of water. Similarly when a base is added to water it takes a proton from water molecules, producing OH^- ions. But these are also present from the dissociation of water.

11-33 $K_w = [H_3O^+][OH^-] = 1.0 \times 10^{-14}$. Therefore, if $(1.0 \times 10^{-12})[OH^-] = 1.0 \times 10^{-14}$, then $[OH^-] = 0.010$ M.

11-35 $[H_3O^+] = 10^{-pH}$

pH = 3.72 then $[H_3O^+] = 1.9 \times 10^{-4}$ M

pH + pOH = 14.00   pOH = 10.28

$[OH^-] = 10^{-pOH}$ $[OH^-] = 5.3 \times 10^{-11}$ M

11-37 When a strong acid is added to water, the hydronium ion (H_3O^+) concentration increases. The hydroxide ion (OH^-) concentration decreases. The pH of the solution also decreases.

11-39 The most acidic water solution will exhibit the smallest pH. Answer (a); the pH of 2.9 for the 0.10 M acetic acid solution indicates that it is the most acidic solution.

11-41 $pH = -\log([H_3O^+])$

$pH = -\log(1.5 \times 10^{-6}) = 5.8$

$pOH = 14 - pH = 8.2$

11-43 $[H_3O^+] = 10^{-pH}$

pH = 3.8 then $[H_3O^+] = 1.6 \times 10^{-4}$M

pH + pOH = 14.00   pOH = 10.2

$[OH^-] = 6.3 \times 10^{-11}$ M

11-45 pH + pOH = 14.00, pH = 11.1, pOH = 2.9

$[OH^-] = 10^{-pOH}$, $[OH^-] = 1.3 \times 10^{-3}$ M

11-47 Strong acids release protons nearly completely while weak acids release protons only to a limited degree. HCl is fully dissociated in water, while acetic acid is not.
Strong bases accept protons most readily while weak bases accept protons only to a limited degree. NaOH is fully dissociated in water, releasing all the hydroxide ions, whereas NH_3 will bind with water but not always release a proton.

11-49

	Acid	K_a
(a)	acetic acid	1.75×10^{-5}
(b)	boric acid	7.3×10^{-10}
(c)	chromic acid	9.6
(d)	formic acid	1.8×10^{-4}
(e)	hydrobromic acid	1×10^9

Both chromic acid and hydrobromic acid qualify as "strong acids" because their $K_a > 1$. The others are classified as weak acids.

11-51   For equal concentrations of weak acids in water solution, the larger the value of Ka the stronger the acid. Cl_2CHCO_2H is the strongest acid of those listed with a $K_a = 5.1 \times 10^{-2}$. Response (d).

11-53   The stronger acid will generate a higher $[H_3O^+]$ concentration in water solution. The correct order of the listed compounds from the weakest to strongest acid is:
$$HOAc < HNO_2 < HF < HOClO$$

11-55 I^-, ClO_4^-, NO_3^- will not act as bases in water since they are the conjugates of strong acids, HI, $HClO_4$, and HNO_3.

11-57   According to the Brønsted model, the stronger acid will transfer a proton, becoming the weaker conjugate base. The stronger base will accept a proton, becoming a weaker conjugate acid. In the reaction of hydrogen chloride with water the stronger proton donor, HCl, transfers the proton to the stronger base, water, creating the weaker conjugate acid, hydronium ion, and the weaker conjugate base, chloride ion.

11-59   According to the Brønsted model, the stronger acid will transfer a proton, becoming the weaker conjugate base. The stronger base will accept a proton, becoming a weaker conjugate acid.
$$HBr(aq) + H_2O(l) \rightarrow H_3O^+(aq) + Br^-(aq)$$
The reaction above indicates that hydrogen bromide is a stronger acid than water because it gives up its proton more readily.

11-61 $HCO_2^-(aq) + H_2O(l) \rightarrow HCO_2H(aq) + OH^-(aq)$
The fact that this reaction will not proceed as written indicates that water is not a stronger acid than formic acid and will not donate a proton to HCO_2^-. The reaction will proceed as:
$$HCO_2H(aq) + OH^-(aq) \rightarrow H_2O(l) + HCO_2^-(aq)$$
This indicates that OH^- is a stronger base than HCO_2^- and will accept a proton from formic acid.

11-63   Since strong acids and bases completely dissociate in water, their apparent strength is based solely upon their concentration in water rather than the value of the dissociation constant for the species. This is because the strength of the H_3O^+ or OH^- produced limits the acid or base strength. This limitation is called the **leveling effect**.

11-65　The H_3O^+ concentration in a strong acid solution depends on the concentration of the solution and not the value of K_a for the acid, because we assume the acid is completely dissociated.

11-67　$AsO_4^{-3} < HAsO_4^{-2} < H_2AsO_4^- < H_3AsO_4$. As the charge gets less negative the acidity increases.

11-69　(d) H_2Te is the strongest acid because the H–X bond length is the longest, meaning that it is also the weakest bond and the easiest to dissociate in water.

11-71　HOI < HOBr < HOCl. The more electronegative the element attached to the oxygen the more likely that electron density will be pulled away from the hydrogen, making it more acidic.

11-73　$CH_3COOH < CH_2ClCOOH < CCl_3COOH$. The more chlorine atoms attached the more likely that electron density will be pulled away from the hydrogen, making it more acidic.

11-75　$$\frac{0.568 \text{ g HCl}}{250 \text{ ml}} \times \frac{1 \text{ mol HCl}}{36.461 \text{ g HCl}} \times \frac{1000 \text{ ml}}{1 \text{ L}} = 0.0623 \frac{\text{mol HCl}}{\text{L}}$$
pH = - log $[H_3O^+]$
pH = 1.2
pOH = 12.8

11-77　(a) $[H_3O^+]$=1.0, pH= 0

(b) $[H_3O^+]$= $\dfrac{0.568 \text{ g HClO}_4}{250 \text{ ml}} \times \dfrac{1 \text{ mol HClO}_4}{100.46 \text{ g HClO}_4} \times \dfrac{1000 \text{ ml}}{1 \text{ L}}$ = 0.0226 M, pH=1.65

(c) $[H_3O^+]$= $\dfrac{0.568 \text{ g HNO}_3}{250 \text{ ml}} \times \dfrac{1 \text{ mol HNO}_3}{63.011 \text{ g HNO}_3} \times \dfrac{1000 \text{ ml}}{1 \text{ L}}$ = 0.0361 M, pH=1.44

(d) $[H_3O^+]$= $\dfrac{1.14 \text{ g HBr}}{500 \text{ ml}} \times \dfrac{1 \text{ mol HBr}}{80.912 \text{ g HBr}} \times \dfrac{1000 \text{ ml}}{1 \text{ L}}$ = 0.0282 M, pH=1.55

11-79　Assuming that HNO_3 is completely dissociated:
pH = - log $[H_3O^+]$ = - log (0.10) = 1.00
Using K_a = 28:
$$\frac{[H_3O^+] [NO_2^-]}{[HNO_3]} = \frac{[\Delta C] [\Delta C]}{[0.10 - \Delta C]} = 28$$
ΔC^2 = 28 ΔC - 2.8 = 0, solving with the quadratic equation,
$$\Delta C = \frac{-28 + \sqrt{28^2 - 4(1)(-2.8)}}{2(1)} = 0.10 \frac{\text{mol } H_3O^+}{\text{L}}$$
pH = -log$[H_3O^+]$ = -log(0.10) = 1.00

11-81 In solutions of strong acids or strong bases, the corresponding hydronium and hydroxide ion concentrations are completely determined, because a strong acid or base fully dissociates in water solution. In solutions of very weak acids or bases, the acid or base does not completely dissociate and therefore more fully associated species are present in addition to dissociated species. For strong acids and bases no associated species are present.

11-83 For solutions of the same weak acid, the solution with the highest molarity also has the highest H_3O^+ ion concentration. $[H_3O^+] = \sqrt{K_a C_{HA}}$. The 0.10 M HOAc solution has the largest H_3O^+ ion concentration of the solutions listed. Response (a).

11-85 For the weak acid formic acid, $[H_3O^+] = (K_a C_{HA})^{1/2}$ and C_{HA} is the initial concentration of HA.
$[H_3O^+] = \sqrt{(1.8 \times 10^{-4})(0.10)} = 4.2 \times 10^{-3}$ M $= [H_3O^+] = [HCO_2^-]$
$[HCO_2H] = 0.100 - 4.2 \times 10^{-3} = 9.6 \times 10^{-2}$ M

11-87 For the weak acid phenol, $[H_3O^+] = (K_a C_{HA})^{1/2}$ and C_{HA} is the initial concentration of HA.
$[H_3O^+] = [PhO^-] = \sqrt{(1.0 \times 10^{-10})(0.0167)} = 1.3 \times 10^{-6}$ M

11-89 Assume that the dissociation of the second proton on ascorbic acid does not contribute to the species in solution.

	$C_6H_6O_6H_2(aq)$	+ $H_2O(l)$	$\rightleftarrows$	$H_3O^+(aq)$	+ $C_6H_6O_6H^-(aq)$
Initial	0.10 M			0	0
Equil	0.10 M-2.8%(0.10M)			2.8%(0.10M)	2.8%(0.10M)

$$K_a = \frac{[C_6H_6O_6H^-][H_3O^+]}{[C_6H_6O_6H_2]} = \frac{(0.0028)^2}{(0.100 - 0.0028)} = 8.1 \times 10^{-5}$$

11-91 The dissociation of pure water has a pH of 7. Adding acid to it, even in small concentrations, will only bring the pH down. The error in this assumption is in forgetting the H_3O^+ present due to the dissociation of water.

11-93 pH$= -\log [H_3O^+] = 4.27$
$$K_a = \frac{[(5.4 \times 10^{-5})(5.4 \times 10^{-5})]}{(0.100)} = 2.91 \times 10^{-8}$$
A 0.10 M solution of HNO_2 would have an $[H_3O^+]$ equal to the concentration of acid, and the pH would be 1.0

11-95 For a conjugate acid - base pair, $K_a K_b = K_w$.
Response (c).

11-97 Assume that the dissociation of the sodium formate is complete.

	$HCO_2^-(aq)$	$+$	$H_2O(l)$	$\rightleftharpoons$	$HCO_2H(aq)$	$+$	$OH^-(aq)$
Initial	0.080 M				0		0
Equil	0.080 - ΔC				ΔC		ΔC

$$K_b = \frac{K_w}{K_a} = \frac{1.0 \times 10^{-14}}{1.8 \times 10^{-4}} = 5.6 \times 10^{-11}$$

$$K_b = \frac{[HCO_2H][OH^-]}{[HCO_2^-]} = 5.6 \times 10^{-11}$$

$$\frac{\Delta C^2}{0.080 - \Delta C} = 5.6 \times 10^{-11}$$

Assume $\Delta C \ll 0.080$

$\Delta C^2 = (0.080)\, 5.6 \times 10^{-11} = 4.5 \times 10^{-12}$

$\Delta C = 2.1 \times 10^{-6}$ M. The assumption is valid.

$[OH^-]=[HCO_2H]=2.1 \times 10^{-6}$ M

$[HCO_2^-]=0.080$ M

11-99 Consider the dissociation of the weak acid HA.

	$OAc^-(aq)$	$+$	$H_2O(l)$	$\rightleftharpoons$	$HOAc(aq)$	$+$	$OH^-(aq)$
Initial	0.756 M				0		0
Equilibrium	0.756 - ΔC				ΔC		ΔC

$$K_b = \frac{K_w}{K_a} = \frac{1.0 \times 10^{-14}}{1.8 \times 10^{-5}} = 5.6 \times 10^{-10}$$

$$K_b = \frac{[HOAc][OH^-]}{[OAc^-]} = 5.6 \times 10^{-10}$$

$$\frac{\Delta C^2}{0.756 - \Delta C} = 5.6 \times 10^{-10}$$

Assume $\Delta C \ll 0.756$

$\Delta C^2 = 4.2 \times 10^{-10}$

$\Delta C = 2.0 \times 10^{-5}$ M. The assumption is valid.

$[OH^-]= 2.0 \times 10^{-5}$ M

pOH = 4.7

pH = 14.00 - 4.7 = 9.3

11-101

	$CH_3NH_2(aq)$	$+$	$H_2O(l)$	$\rightleftharpoons$	$CH_3NH_3^+(aq)$	$+$	$OH^-(aq)$
Equil	93.2%(0.10M)				6.8%(0.10M)		6.8%(0.10M)

$$K_b = \frac{[CH_3NH_3^+][OH^-]}{[CH_3NH_2]} = \frac{(0.0068)^2}{0.093} = 5.0 \times 10^{-4}$$

Methylamine is a stronger base than aqueous ammonia.

11-103 There are two ways to calculate the pH of a 0.10 M solution of methylamine given the information contained in problem 101.

1. The $[OH^-]$ is $\dfrac{6.8}{100}(0.10) = 0.0068$

from which pOH is 2.17 and pH = 14.00-2.17 = 11.83.

2. K_b was calculated to be 5.0 x 10^{-4}.

$$K_b = \frac{[CH_3NH_3^+][OH^-]}{0.10-\Delta C} = 5.0 \times 10^{-4}$$

$$\frac{\Delta C^2}{0.10 - \Delta C} = 5.0 \times 10^{-4}$$

Solving the quadratic equation, we get $\Delta C=[OH^-]=0.0068$; pH=11.8.

11-105

$$OAc^-(aq) \quad + \quad H_2O(l) \quad \rightleftarrows \quad HOAc(aq) \quad + \quad OH^-(aq)$$

Initial	0.032 M		0	0
Equilibrium	0.032 - ΔC		ΔC	ΔC

The concentration of AcO^- ion $= 0.016 \dfrac{mol\ Ca(OAc)_2}{liter} \times \dfrac{2\ mol\ OAc^-}{1\ mol\ Ca(OAc)_2} = 0.032\ M.$

$$K_b = \frac{K_w}{K_a} = \frac{1.0 \times 10^{-14}}{1.8 \times 10^{-5}} = 5.6 \times 10^{-10}$$

$$K_b = \frac{[HOAc][OH^-]}{[OAc^-]} = 5.6 \times 10^{-10}$$

$$\frac{\Delta c^2}{0.032 - \Delta c} = 5.6 \times 10^{-10}$$

Assume $\Delta C << 0.032$

$\Delta C^2 = 1.8 \times 10^{-11}$ $\Delta C = 4.2 \times 10^{-6}$ M. The assumption is valid.

$[OH^-] = 4.2 \times 10^{-6}$ M   pOH = 5.4

pH = 8.6

11-107   (a) NH_4Cl. NH_4^+ is a weak acid, whereas NO_3^- and F^- are conjugates of weak acids.

11-109   (a)NaH_2PO_4 (b)H_2CO_3 (c) $NaHSO_4$ (d) HNO_2

11-111   For the weak acid HF, $[H_3O^+] = (K_aC_{HA})^{1/2}$ where C_{HA} is the initial concentration of HA.

$C_{HA} = 0.50$ M

$$[H_3O^+] = \sqrt{(7.2\times10^{-4})(0.50}= 1.9 \times 10^{-2}\ M,\ pH= 1.72$$

$$K_a = \frac{[F^-][H_3O^+]}{[HF]},\ [H_3O^+] = \frac{K_a[HF]}{[F^-]} = \frac{(7.2\times10^{-4})(0.50)}{0.50} = 7.2 \times 10^{-4},\ pH=3.1$$

11-113 $K_a = \dfrac{[NH_3][H_3O^+]}{[NH_4^+]}$, $[H_3O^+] = \dfrac{K_a[NH_4^+]}{[NH_3]} = \dfrac{(5.6\times10^{-10})(0.10)}{0.1} = 5.6 \times 10^{-10},\ pH=9.3$

11-115 A **buffer** is composed of both the weak acid or base and its conjugate form at appreciable concentration. Addition of an acid consumes some of the conjugate base form buffering the influence of the addition. For addition of a base, some of the conjugate acid form of the buffering pair is consumed. The net result is that the hydronium ion concentration in the solution does not change very much when either an acid or a base is added to the solution.

11-117 To be an effective buffer, we seek systems that have a conjugate pair of substances related to each other by transfer of a proton. Neither should be exceptionally strong. Response (e).

11-119 Response (d) is a basic buffer.

11-121 We can rearrange the relationship $K_a \times \dfrac{[HA]}{[A^-]} = [H_3O^+]$ to get $\dfrac{[HA]}{[A^-]} = \dfrac{[H_3O^+]}{K_a}$.

A pH=5.6 means that the $[H_3O^+] = 2.51 \times 10^{-6}$.

For acetic acid $\dfrac{[HA]}{[A^-]} = \dfrac{2.51 \times 10^{-6}}{1.75 \times 10^{-5}} = 0.143$

For chlorous acid $\dfrac{[HA]}{[A^-]} = \dfrac{2.51 \times 10^{-6}}{1.1 \times 10^{-2}} = 0.00023$

For formic acid $\dfrac{[HA]}{[A^-]} = \dfrac{2.51 \times 10^{-6}}{1.8 \times 10^{-4}} = 0.014$

11-123 The K_a for the HCO_3^- ion is 4.7×10^{-11}. Using $K_a \times \dfrac{[HA]}{[A^-]} = [H_3O^+]$ we find that when the $[HA]=[A^-]$ that $[H_3O^+] = K_a$. Thus for this system $[H_3O^+] = 4.7 \times 10^{-11}$. pH = 10.3.

11-125 (a), (c), and (d) will all go nearly to completion.

11-127 pH= 7.0 Equal moles of a strong acid and base have been added. They will neutralize each other.

11-129 (a) $HOBr(aq) + OH^-(aq) \rightleftarrows H_2O(l) + OBr^-(aq)$

(b) $HOAc(aq) + NH_3(aq) \rightleftarrows OAc^-(aq) + NH_4^+(aq)$

(c) $H_3O^+(aq) + OH^-(aq) \rightleftarrows 2\ H_2O(l)$

(d) $2\ HCOOH(aq) + Ba(OH)_2(s) \rightleftarrows 2\ HCOO^-(aq) + 2\ H_2O(l) + Ba^+(aq)$

Reactions (a), (c) and (d) will go to completion.

11-131 A **titration** is when a known quantity of one reagent is added to a second solution until the end point has been reached. The **end point** is when an indication is given that the equivalence point of the titration has been reached. The **equivalence point** in a titration is when the known amount of material added by titration is equal to the amount of material already present in the solution. That is, equivalent amounts of each substance have been added to the solution. The end point is determined by use of an **indicator**, which changes color when the end point has been reached.

11-133 The titration curve for the titration of the weak base NH_3 with strong acid would be as follows:

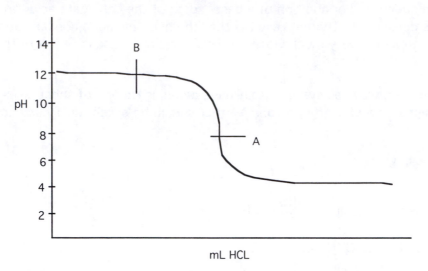

mL HCL

Point A is the equivalence point of the titration. Equal molar amounts of acid and base have reacted and the curve drops steeply.

Point B is the half way point in the titration. At this point in the titration, the amounts of NH_3 and NH_4^+ are equal and the $[OH^-] = K_b$ for the base.

11-135 Referring to table 11.7 the following colors will be observed at pH=10.0:
(a) methyl orange: yellow
(b) cresol red: red
(c) phenolphthalein: pink
(d) alizarin yellow: yellow

11-137 By definition:
$[H_3O^+] = 10^{-pH}$
The rain in Scotland had $[H_3O^+] = 10^{-2.4}$ mol/L = 0.004 M.
The acetic acid solution had $[H_3O^+] = 10^{-2.9}$ mol/L = 0.001M.

11-139 (a) False, each HCl that dissociates produces an H_3O^+ and a Cl^-.
(b) True
(c) True
(d) False, since HCl is a strong acid, it completely dissociates.

11-141 (a) The hydrogen attached to the more electronegative atom, oxygen, is acidic.

(b)

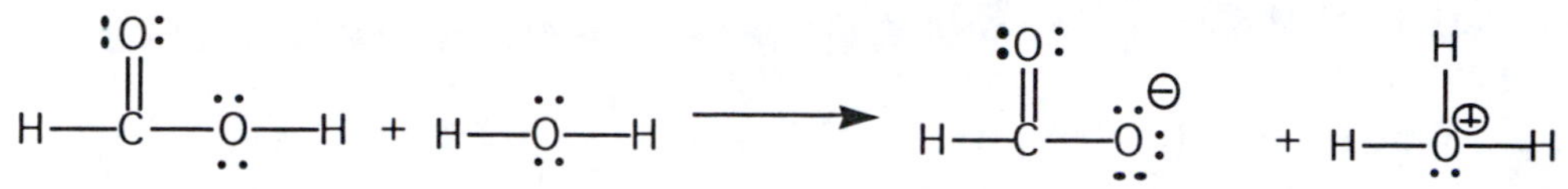

(c)

	HCO_2H	+	$H_2O(l)$	$\rightleftharpoons$	$H_3O^+(aq)$	+	$HCO_2^-(aq)$
Initial	0.10 M				0		0
Equil	0.10 - ΔC				ΔC		ΔC

since pH = 2.37, $10^{-2.37}$ = $[H_3O^+]$ = ΔC = 4.3 x 10^{-3}

$$K_a = \frac{[HCO_2^-][H_3O^+]}{[HCO_2H]} = \frac{(4.3 \times 10^{-3})(4.3 \times 10^{-3})}{(0.10)} = 1.8 \times 10^{-4}$$

(d) The pH of the solution should be less than 2.37 because the addition of the Cl atom should cause an increase in the strength of the acid.

11-143 A list in order of strongest acid to strongest base will also be in order of increasing pH.
$HBr < HOBrO_2 < HOBrO < NaBr < NaBrO_3 < NaBrO_2$

11-145 (a) A dilute solution of a strong acid would consist of A^- and H_3O^+ in small amounts. Figure (iii) matches this description.
(b) A concentrated solution of a weak acid would consist of a majority of HA and some H_3O^+ and A^-. Figure (i) matches this description.
(c) A buffer solution would consist of equal amounts of HA and A^- with some H_3O^+. Figure (iv) matches this description.

11-147 Conjugate base: (c) Conjugate acid: (a)

11-149 (a) $Al_2O_3(s) + 6\ HCl(aq) \rightarrow 2\ AlCl_3(aq) + 3\ H_2O(l)$
(b) $CaO(s) + H_2SO_4(aq) \rightarrow CaSO_4(aq) + H_2O(l)$
(c) $Na_2O(s) + H_2O(l) \rightarrow 2Na^+(aq) + 2HO^-(aq)$
(d) $MgCO_3(s) + 2\ HCl(aq) \rightarrow H_2CO_3(aq) + MgCl_2(aq)$
(e) $3\ NaOH(s) + H_3PO_4(aq) \rightarrow 3\ H_2O(aq) + Na_3PO_4(aq)$

105

11-151 A. a) $HBr(aq) + H_2O(l) \rightarrow H_3O^+(aq) + Br^-(aq)$

b) $[H_3O^+] = 1.0\ M$

c) $pH = 0$

d) $[OH^-] = 1.0 \times 10^{-14}$

e) $Br^-(aq)$

B. a) $NH_3(aq) + H_2O(l) \rightarrow NH_4^+(aq) + OH^-(aq)$

b) $K_b = \dfrac{[NH_4^+][OH^-]}{[NH_3]} = \dfrac{[OH^-]^2}{[NH_3]} = 1.8 \times 10^{-5}$,

$[OH^-] = \sqrt{K_b \times [NH_3]} = 4.2 \times 10^{-3}\ M$

c) $[NH_3] = 1.0 - 4.2 \times 10^{-3} = 1.0\ M$

d) $[NH_4^+] = 4.2 \times 10^{-3}\ M$

e) $pH = 14 - pOH = 11.6$

C. a) $[Br^-] = 1.0\ M$

b) $Br^-(aq) + H_2O(l) \rightarrow HBr(aq) + OH^-(aq)$ The equilibrium constant will be low, because Br^- is a conjugate base of a weak acid.

c) $pH = 7.0$

11-153 (a) $HNO_2(aq) + H_2O(l) \rightarrow NO_2^-(aq) + H_3O^+(aq)$

(b) $K_a = \dfrac{[NO_2^-][H_3O^+]}{[HNO_2]} = \dfrac{\Delta C \cdot \Delta C}{1.00} = 5.1 \times 10^{-4}$

$\Delta C = 2.3 \times 10^{-2}\ M.$

$pH = 1.65$

(c) $K_b = \dfrac{K_w}{K_a} = 1.96 \times 10^{-11} = \dfrac{[HNO_2][OH^-]}{[NO_2^-]} = \dfrac{\Delta C \cdot \Delta C}{1.00}$

$\Delta C = 4.4 \times 10^{-6}\ M$

$pOH = 5.35, pH = 8.65$

d) $pH = 0$, strong acid

e) Diagram (a), because the nitric acid will completely dissociate.

11-155 a) $pH = 1.0$, strong acid

b) $pH = 7.0$, conjugate base of strong acid

c) $1 < pH < 7$, this is a weak acid

d) $7 < pH < 14$, this is a weak base (conjugate of the weak acid)

e) $pH = 13$, strong base

11-157 Increasing acidity

KOH   strong base

$KClO_3$ conjugate base of weak acid

KI   conjugate base of strong acid

NH_4ClO_4 weak acid and conjugate base of strong acid

$HClO_3$ weak acid

$HClO_4$ strong acid

11-159 At the start of the titration, $K_a = \dfrac{[OAc^-][H_3O^+]}{[HOAc]} = \dfrac{\Delta C \cdot \Delta C}{0.100} = 1.8 \times 10^{-5}$

$\Delta C = 1.3 \times 10^{-3}$ M

pH = 2.87

After adding 20.0 mL of 0.10 M NaOH, there will be 0.003 moles of HOAc left in 70.0 mls, the [HOAc]= 0.0429 M.

There will also be 0.002 moles of OAc⁻ in the 70.0 mls, the [OAc⁻]= 0.0286 M.

$K_a = \dfrac{[OAc^-][H_3O^+]}{[HOAc]} = \dfrac{0.0286 \cdot [H_3O^+]}{0.0429} = 1.8 \times 10^{-5}$

$[H_3O^+] = 2.7 \times 10^{-5}$ M

pH=4.57.

At the equivalence point when 50 mL of 0.10 M NaOH has been added, there will be 0.0050 moles of OAc⁻ in 100.0 mls of solution, [OAc⁻]= 0.050 M

$K_b = \dfrac{K_w}{K_a} = 5.56 \times 10^{-10} = \dfrac{[HOAc][OH^-]}{[OAc^-]} = \dfrac{\Delta C \cdot \Delta C}{0.050}$

$\Delta C = 5.3 \times 10^{-6}$ M

pOH=5.28

pH=8.72

After adding 60.0 ml of the base the base concentration remaining is 0.001 moles of OH⁻ in 110 ml, [OH⁻]=0.0091 M. This will also be a greater contribution to the pH than the weak base, OAc⁻ , which is also in the solution.

pH= 12.0.

11-161 (b) HSO_4^- is the conjugate base of the strong acid sulfuric acid. H_3O^+ is a strong acid. The remainder of the ions are conjugate bases of weak acids.

Chapter 11 Special Topics

11A-1 $K_{a1} = \dfrac{[HCO_3^-][H_3O^+]}{[H_2CO_3]} = 4.5 \times 10^{-7}$ $K_{a2} = \dfrac{[CO_3^{2-}][H_3O^+]}{[HCO_3^-]} = 4.7 \times 10^{-11}$

Assume $[H_2CO_3]$ at equilibrium = 0.100 M.

Assume $[HCO_3^-] = [H_3O^+] = \Delta C$

$\dfrac{\Delta C^2}{0.100} = 4.5 \times 10^{-7}$, $\Delta C^2 = 4.5 \times 10^{-8}$, $\Delta C = 2.1 \times 10^{-4}$ M

$[HCO_3^-] = [H_3O^+] = 2.1 \times 10^{-4}$ M

$\dfrac{[CO_3^{2-}][H_3O^+]}{[HCO_3^-]} = 4.7 \times 10^{-11}$

$\dfrac{[CO_3^{2-}]\left(2.1 \times 10^{-4}\right)}{\left(2.1 \times 10^{-4}\right)} = 4.7 \times 10^{-11}$ M

$[CO_3^{2-}] = 4.7 \times 10^{-11}$ M

11A-3 Check the assumption that $[H_3O^+] \cong [HM^-]$

$[H_3O^+]$ and $[HM^-]$ are $\dfrac{1.87 \times 10^{-3}}{0.250} \times 100 = 0.75\%$ of the initial value of H_2M. Therefore the assumption that $[H_2M] = H_2M$ initial is valid.

The concentration of $M^{2-} = 2.1 \times 10^{-8}$ M. This is about 89,000 times smaller than $[HM^-]$ and $[H_3O^+]$. So the assumption that nearly all of the HM^- and H_3O^+ come from the first dissociation is valid.

11A-5 (a) 9.3×10^{-2} M < 1.9 M   no

(b) 9.3×10^{-2} M < 1.9 M   yes

(c) 9.3×10^{-2} M $= 9.3 \times 10^{-2}$ M   no

(d) 9.3×10^{-2} M $= 9.3 \times 10^{-2}$ M   yes

(e) 9.3×10^{-2} M $= 9.3 \times 10^{-2}$ M   no

Responses (b) and (d)

11A-7   Assume stepwise dissociation of oxalic acid,

$$K_{a1} = \frac{[H_3O^+][HC_2O_4^-]}{[H_2C_2O_4]} = 5.4 \times 10^{-2}$$

$$K_{a2} = \frac{[H_3O^+][C_2O_4^{2-}]}{[HC_2O_4^-]} = 5.4 \times 10^{-5}$$

Working with the first dissociation,

$$\frac{\Delta C^2}{1.25 - \Delta C} = 5.4 \times 10^{-2}$$

First approximation: $\dfrac{(\Delta C')^2}{1.25} = 5.4 \times 10^{-2}$ $\Delta C' = 0.260$

Second approximation: $\dfrac{(\Delta C'')^2}{1.25 - 0.260} = 5.4 \times 10^{-2}$ $\Delta C'' = 0.231$

Third approximation: $\dfrac{(\Delta C''')^2}{1.25 - 0.260} = 4.5 \times 10^{-3}$ $\Delta C''' = 0.235$

Check, $\dfrac{(0.235)^2}{(1.25 - 0.235)} = 5.44 \times 10^{-2}$ M which agrees with the value given.

$\Delta C = 0.24$ M $= [H_3O^+] = [HC_2O_4^-]$, and $[H_2C_2O_4] = 1.01$ M

11A-9 $C_2O_4^{2-}(aq) + H_2O(l) \rightleftarrows HC_2O_4^-(aq) + OH^-(aq)$

$$K_{b1} = \frac{K_w}{K_{a2}} = \frac{1.0 \times 10^{-14}}{5.4 \times 10^{-5}} = 1.9 \times 10^{-10}$$

$HC_2O_4^-(aq) + H_2O(l) \rightleftarrows H_2C_2O_4^-(aq) + OH^-(aq)$

$$K_{b2} = \frac{K_w}{K_{a1}} = \frac{1.0 \times 10^{-14}}{5.4 \times 10^{-2}} = 1.9 \times 10^{-13}$$

Working with K_{b1},

Assume $[OH^-] \cong [HC_2O_4^-] = \Delta C$ and $[C_2O_4^{2-}] \cong C_{Na_2C_2O_4}$

$$\frac{\Delta C^2}{0.028 - \Delta C} = 1.9 \times 10^{-10}$$

$\Delta C^2 = (0.028)(1.9 \times 10^{-10}) = 5.3 \times 10^{-12}$

$\Delta C = 2.3 \times 10^{-6}$ M $= [OH^-] = [HC_2O_4^-]$

$[C_2O_4^{2-}] = 0.028$ M

pOH $= 5.6$

pH $= 8.4$

11A-11   When solid $NaHCO_3$ is put into water, it ionizes to give Na^+ and HCO_3^- ions. HCO_3^- ion is a stronger base than it is an acid. The HCO_3^- ions will accept a proton. The only substance in solution that can donate a proton to HCO_3^- is the water molecule. This leaves an excess of OH^- ions in solution. This is why the solution is basic.

11A-13   0.10 M H_3PO_4 $K_{a1} = 7.1 \times 10^{-3}$

H_3PO_4 is acidic $[H_3O^+] = \Delta C$

$\Delta C^2 = 7.1 \times 10^{-4}$

$\Delta C = 2.7 \times 10^{-2} = [H_3O^+]$ pH = 1.6

0.10 M $NaHPO_4$

The $H_2PO_4^-$ ion is acidic $K_{a2} = 6.3 \times 10^{-8}$

$[H_3O^+] = \Delta C$

$\Delta C^2 = 6.3 \times 10^{-9}$ and $\Delta C = 7.9 \times 10^{-5}$ pH = 4.1

0.10 M Na_2HPO_4

The HPO_4^{2-} ion is basic $K_{b2} = 1.6 \times 10^{-7}$

$[OH^-] = \Delta C$

$\Delta C^2 = 1.6 \times 10^{-8}$ and $\Delta C = 1.3 \times 10^{-4}$

pOH = 3.9 and pH = 10.1

0.10 M Na_3PO_4

The PO_4^{3-} ion is basic $K_{b1} = \dfrac{K_w}{K_{a3}} = \dfrac{1.0 \times 10^{-14}}{4.2 \times 10^{-13}} = 0.024$

$[OH^-] = \Delta C$

$\Delta C^2 = 2.4 \times 10^{-3}$ and $\Delta C = 4.9 \times 10^{-2}$

pOH = 1.3 and pH = 12.7

11A-15 $K_{b2} = \dfrac{K_w}{K_{a1}} = \dfrac{1 \times 10^{-14}}{1.7 \times 10^{-2}} = 5.9 \times 10^{-13}$

$K_{a2} > K_{b2}$, the solution of $NaHSO_3$ will be acidic.

12-1   **Oxidation** occurs when an atom or molecule loses electrons in the course of a reaction. **Reduction** occurs when an atom or molecule gains electrons in the course of a reaction.

12-3   (a) reduction
       (b) oxidation
       (c) reduction
       (d) oxidation

12-5   +3 in all the compounds.

12-7   Taking the oxidation state of H as + 1 and that of O as - 2, the order of increasing oxidation state of the carbon atoms is

CH_4 (-4) < H_3COH (-2) < C and H_2CO (0) < CO (+2) < CO_2 (+4)

12-9
(a)	Na_2S	-2	(e)	SF_6	+6
(b)	MgS	-2	(f)	$BaSO_3$	+4
(c)	CS_2	-2	(g)	$NaHSO_4$	+6
(d)	H_2SO_4	+6	(h)	SCl_2	+2

12-11   The oxidation number of the chlorine increases from +1 in HClO to +7 in $HClO_4$. As oxidation number increases, the strength of the acid increases.

12-13   (a) $Mg(s) + 2 H_3O^+(aq) \rightleftarrows Mg^{2+}(aq) + Cr(s) H_2(g) + 2 H_2O(l)$

12-15   This is not a redox reaction.

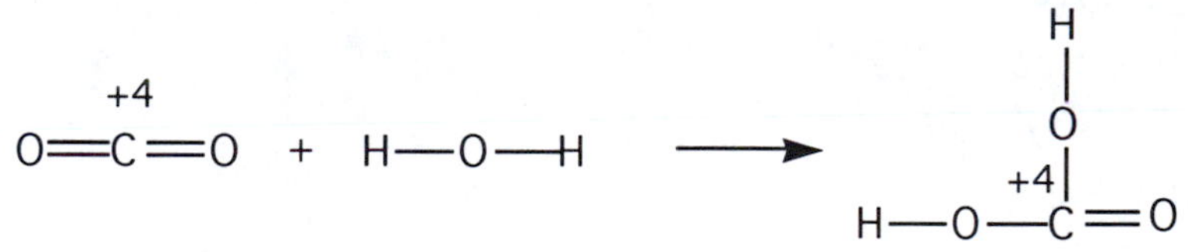

$$\overset{+4}{O}=C=O \ + \ H-O-H \ \longrightarrow \ H-O-\overset{+4}{C}=O \ (OH)$$

12-17   In the reaction below the sulfur is oxidized (+4 → +6) and
        the oxygen gas is reduced (0 → -2)

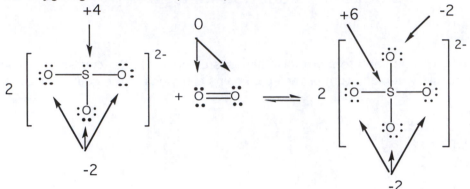

12-19   Each carbon in ethylene has been
        reduced from an oxidation state of -2
        to an oxidation state of -3 in ethane.
        This is equivalent to a gain of two
        electrons. The hydrogenation reaction
        is a reduction.

12-21   Salt bridges contain a mixture of anions and cations. During oxidation and reduction
        reactions, these ions are able to flow from the salt bridge into the half-reaction solutions
        to balance the charge created by the oxidation and reduction. If the salt bridge is
        removed, there will be a buildup of positive charge at the anode, and negative charge at
        the cathode and the reaction will stop.

12-23   As a compound moves to the cathode and is reduced by accepting an electron, the
        solution becomes more negative. Since charged solutions are unstable, cations move from
        the salt bridge to maintain an electrically neutral solution.

12-25   (a) $Al(s) \rightleftarrows Al^{3+}(aq) + 3\ e^-$
             $3\ e^- + Cr^{3+}(aq) \rightleftarrows Cr(s)$
        (b) $2\ Fe^{2+}(aq) \rightleftarrows 2\ Fe^{3+}(aq) + 2\ e^-$
             $2\ e^- + I_2(aq) \rightleftarrows 2\ I^-(aq)$
        (c) $Cr(s) \rightleftarrows Cr^{3+}(aq) + + 3\ e^-$
             $3\ e^- + 3\ Fe^{3+}(aq) \rightleftarrows 3\ Fe^{2+}(aq)$
        (d) $Zn(s) \rightleftarrows Zn^{2+}(aq) + 2\ e^-$
             $2\ e^- + 2\ H^+(aq) \rightleftarrows H_2(g)$

12-27   The sign of the cell potential tells the direction in which the reaction must shift to reach
        equilibrium. Oxidation-reduction reactions that have a positive overall cell potential are
        spontaneous as written. Those that have a negative overall cell potential are not
        spontaneous as written. An overall cell potential of zero indicates a system at equilibrium.

12-29   The sign of the standard state potential for an oxidation-reduction reaction must change
        when the direction in which the reaction is written is reversed. The magnitude remains the
        same.

12-31   (a) The Al is oxidized and Cr^{3+} is reduced. So the Al is the reducing agent while the Cr^{3+} is the oxidizing agent.
(b) One of the Cr^{2+} is reduced to Cr and the two others are oxidized to Cr^{3+}. In this case the Cr^{2+} is both the oxidizing and reducing agent.
(c) The Fe is oxidized and Cr^{3+} is reduced, so the Fe is the reducing agent and Cr^{3+} is the oxidizing agent.
(d) The H_2 is being oxidized and Cr^{3+} is reduced, so the H_2 is the reducing agent and Cr^{3+} is the oxidizing agent.

12-33   Yes, because when it oxidizes another substance it will gain electrons, in the +4 oxidation state carbon will not lose any more electrons.

12-35   An oxidizing agent must itself be reduced in the process. Chloride ion (a) and zinc metal (d) cannot be further reduced. The calcium ion of calcium hydride (e) does not become reduced except under very select conditions. Only bromine (b) and iron(3+) (c) are expected to be oxidizing agents.

12-37   (c) H_2O_2 will not be a reducing agent. All of the other substances listed can be either oxidizing or reducing agents under the proper conditions.

12-39   When an atom is in the lowest, or most negative, oxidation state it cannot take in any more electrons and will therefore not be an oxidizing agent. So it can only act as a reducing agent.

12-41   When using a table of relative reactivity such as Table 12.1, there is a preferred direction for combination of spontaneous reactions. In this situation, spontaneous reactions occur in a counter clockwise manner.   A reaction higher in the table proceeds from right to left when coupled with a reaction farther down in the table (which must proceed from left to right).
(a)   The reverse reaction should occur. So the reaction will not occur as written.
(b)   The reverse reaction should occur. So the reaction will not occur as written.
(c)   This reaction should occur as written.

12-43   (a) $Cr^{3+}(aq)$, $Fe^{3+}(aq)$ (or $Fe^{2+}(aq)$)
(b) $Mg^{2+}(aq)$, $Cr^{2+}(aq)$
(c)   These ions will not react.

12-45   The better oxidizing agents are found at the bottom of Table 12.1. Fluorine reacting in acidic solution, response (d), is the strongest oxidizing agent of the group.

12-47   Set up a galvanic salt bridge cell with a copper metal electrode in a 1M Cu^{2+} solution and an iron metal electrode in a 1M Fe^{3+} solution. Allow the reaction to occur. Observe the electrodes to see at which electrode additional metal is deposited. The electrode at which metal is being deposited is being reduced, thus the other solution is a better reducing agent. By examining the reduction potentials we can determine that the copper electrode will undergo the reduction reaction, depositing copper metal on the electrode surface. The iron metal will be the reducing agent. Therefore, iron metal is a better reducing agent than copper.

12-49   The strongest oxidizing agent will be the substance with the most positive standard reduction potential. Response (e).

12-51   The better reducing agent will have the most positive standard reduction potential.
(a)   K
(b)   Sn
(c) Au^+

12-53   The more positive the standard cell potential for an oxidation-reduction reaction, the more likely the reaction is to proceed in the direction written. Since the Fe^{2+} oxidation is common to each reaction, the magnitude of the standard cell potential is a direct measure of the relative strength of the oxidizing agents, $Cl_2 > Br_2 > H_2O_2 > Fe^{3+} > I_2$.

12-55 $E°_{overall} = E°_{ox} + E°_{red}$
(a) $E°_{overall} = -1.491\ V + 0.770 = -0.721\ V$
(b) $-0.3402\ V + 0.96\ V = 0.62\ V$
Reaction (b) will occur as written. Reaction (a) will not.

12-57 $E°_{overall} = E°_{ox} + E°_{red} = 0.036\ V - 1.706\ V = -1.67\ V$
The reaction should not occur spontaneously as written.

12-59 $E°_{overall} = E°_{ox} + E°_{red} = -0.158\ V + 0.522\ V = 0.364\ V$
The reaction should occur spontaneously as written.

12-61

2 (	HNO_3	+	$3H^+$	+	$3e^-$	$\rightarrow$	NO	+	$2\ H_2O$	)	0.96 V
3 (					Zn	$\rightarrow$	Zn^{2+}	+	$2e-$	)	+0.76 V
	$2\ HNO_3$	+	$6\ H^+$	+	3 Zn	$\rightarrow$	2 NO	+	$4\ H_2O$	+ $3\ Zn^{2+}$	+1.72 V
	$2H^+$	+	$2e^-$	$\rightarrow$	H_2				0.00 V		
	Zn	$\rightarrow$	Zn^{2+}	+	$2e^-$				0.76 V		
	Zn	+	$2\ H^+$	$\rightarrow$	Zn^{2+}	+	H_2	+0.76 V			

Both reactions have positive standard potentials and would be expected to occur.

12-63   When a lead acid battery is discharged PbO_2 is reduced to $PbSO_4$. When it is charged up again the reverse reaction is driven through electrolysis.

12-65   A fuel cell is an electrochemical cell in which a controlled oxidation of a fuel produces an electrical voltage in the same way a battery does.

12-67   The standard-state cell potential, E°, is defined for a single set of conditions when the solution concentrations are 1.0 M and the partial pressures of all gases are at 1.0 atmosphere. The cell potential, E, refers to potentials derived for all other conditions.

12-69   When the cell potential, E, is zero, the reacting system is at equilibrium. When E° is equal to zero, the standard-state cell potential for the reaction is identical to that which we have defined for the standard hydrogen electrode potential.

12-71 $E_{cell} = E° - \dfrac{0.02569}{n}\ln Q_c$

$E_{cell} = E°_{ox} + E°_{red} = +0.036\ V + (-1.706\ V) = -1.67\ V$

$E_{cell} = -1.67\ V - \dfrac{0.02569}{3}\ln\dfrac{(Fe^{3+})}{(Al^{3+})}$

$= -1.67\ V - \dfrac{0.02569}{3}\ln\left(\dfrac{1.0\times10^{-4}}{1.2}\right) = -1.68\ V$

at $T=298K$, $K = e^{\frac{nFE°}{RT}}$, where $n=3$,

$K = e^{-196} = 5.7\times10^{-86}$

This reaction will not proceed.

12-73 For the Daniell cell, $E° = +0.76\ V + 0.34\ V = 1.10\ V$

$E_{cell} = 1.10\ V - \dfrac{0.02569}{2}\ln\dfrac{(Zn^{2+})}{(Cu^{2+})}$

How to get the value of Q_c:

Initially we have 1.0M Cu^{2+} and 1.0M Zn^{2+}. When the reaction has used up 99% of Cu^{2+} then there is only 0.01M Cu^{2+} remaining, and the Zn^{2+} has increased to 1.99M. This

makes $Q_c = \dfrac{(Zn^{2+})}{(Cu^{2+})} = \dfrac{1.99}{0.01} = 199$

(a) $Q_c = 199$ $E_{cell} = 1.03\ V$
(b) $Q_c = 1999$ $E_{cell} = 1.00\ V$
(c) $Q_c = 1999999$ $E_{cell} = 0.91\ V$
(d) $Q_c = 199999999$ $E_{cell} = 0.86\ V$
(e) $E_{cell} = 0.0\ V$

12-75 $E_{cell} = E°_{cell} - \dfrac{RT}{nF}\ln Q_c$

An increase in temperature would lead to a decrease in E_{cell}.

12-77 At equilibrium, $\ln K = \dfrac{nE°}{0.02569} = \dfrac{(2)(1.10)}{0.02569} = 85.7$

$K = e^{85.7} = 2\times10^{37}$

The very high value of the equilibrium constant, showing a very large driving force for product formation, justifies the use of the single arrow in the reaction equation.

12-79 The deposition of silver metal from solution can be measured. If the number of moles of metal deposited per unit time is determined, Faraday's law provides a link to the amount of the electric current.

12-81 $1.0\ mol\ Na \times \dfrac{1\ mol\ e^-}{1\ mol\ Na} \times \dfrac{96485\ C}{1\ mol\ e^-} \times \dfrac{amp\ s}{1\ C} \times \dfrac{1}{1.0\ amp} = 96485\ s$

12-83 Electrolysis of LiBr aqueous solution yields H_2 and Br_2.

$$1.0 \text{ h} \times \frac{3600 \text{ s}}{1 \text{ h}} \times \frac{1 \text{ C}}{\text{amp s}} \times 2.5 \text{ amp} \times \frac{1 \text{ mol e}^-}{96485 \text{ C}} \times \frac{2.0158 \text{ g } H_2}{1 \text{ mol } H_2} \times \frac{1 \text{ mol } H_2}{2 \text{ mol e}^-} = 0.094 \text{ g } H_2$$

$$1.0 \text{ h} \times \frac{3600 \text{ s}}{1 \text{ h}} \times \frac{1 \text{ C}}{\text{amp s}} \times 2.50 \text{ amp} \times \frac{1 \text{ mol e}^-}{96485 \text{ C}} \times \frac{1 \text{ mol } Br_2}{2 \text{ mol e}^-} \times \frac{159.808 \text{ g } Br_2}{1 \text{ mol } Br_2} = 7.4 \text{ g } Br_2$$

12-85

$$2.5 \text{ h} \times \frac{3600 \text{ s}}{1 \text{ h}} \times \frac{1 \text{ C}}{\text{amp s}} \times 4.5 \text{ amp} \times \frac{1 \text{ mol e}^-}{96485 \text{ C}} \times \frac{1 \text{ mol Cu}}{2 \text{ mol e}^-} = 0.21 \text{ mol Cu}$$

$$0.21 \text{ mol Cu} \times \frac{63.546 \text{ g Cu}}{\text{mol Cu}} = 13 \text{ g of copper}$$

12-87

$$1.00 \text{ g Ag} \times \frac{1.00 \text{ g Ag}}{107.87 \text{ g Ag}} \times \frac{1 \text{ mol e}^-}{1 \text{ mol Ag}} \times \frac{96485 \text{ C}}{1 \text{ mol e}^-} = 894.48 \text{ C}$$

$$894.48 \text{ C} \times \frac{1 \text{ mol e}^-}{96485 \text{ C}} \times \frac{1 \text{ mol } I_2}{2 \text{ mol e}^-} \times \frac{253.80 \text{ g } I_2}{1 \text{ mol } I_2} = 1.18 \text{ g } I_2$$

An alternative method of working the problem is by looking at the ratios of the molecular masses and moles of electrons.

$$\frac{1.00 \text{ g Ag}}{107.87 \text{ g Ag}} \times \frac{1 \text{ mol e}^-}{1 \text{ mol Ag}} \times \frac{1 \text{ mol } I_2}{2 \text{ mol e}^-} \times \frac{253.80 \text{ g } I_2}{1 \text{ mol } I_2} = 1.18 \text{ g } I_2$$

12-89 The number of coulombs involved =

$$20.0 \text{ min} \times \frac{60 \text{ s}}{\text{min}} \times \frac{1 \text{ C}}{\text{amp s}} \times 10.0 \text{ amp} = 1.20 \times 10^4 \text{ C}$$

This corresponds to $1.20 \times 10^4 \text{ C} \times \dfrac{1 \text{ mol e}^-}{96485 \text{ C}} = 0.124 \text{ mol e}^-$

(a) $Zn^{2+} + 2 \text{ e}^- \rightarrow Zn$

$$0.124 \text{ mol e}^- \times \frac{1 \text{ mol Zn}}{2 \text{ mol e}^-} \times \frac{65.39 \text{ g Zn}}{1 \text{ mol Zn}} = 4.05 \text{ g Zn}$$

(b) $Zn^{2+} + 2 \text{ e}^- \rightarrow Zn$

$$0.124 \text{ mol e}^- \times \frac{1 \text{ mol Zn}}{2 \text{ mol e}^-} \times \frac{65.39 \text{ g Zn}}{1 \text{ mol Zn}} = 4.05 \text{ g Zn}$$

(c) $W^{6+} + 6 \text{ e}^- \rightarrow W$

$$0.124 \text{ mol e}^- \times \frac{1 \text{ mol W}}{6 \text{ mol e}^-} \times \frac{183.85 \text{ g W}}{1 \text{ mol W}} = 3.80 \text{ g W}$$

(d) $Sc^{3+} + 3 \text{ e}^- \rightarrow Sc$

$$0.124 \text{ mol e}^- \times \frac{1 \text{ mol Sc}}{3 \text{ mol e}^-} \times \frac{44.956 \text{ g Sc}}{1 \text{ mol Hf}} = 1.86 \text{ g Sc}$$

(e) $Hf^{4+} + 4 \text{ e}^- \rightarrow Hf$

$$0.124 \text{ mol e}^- \times \frac{1 \text{ mol Hf}}{4 \text{ mol e}^-} \times \frac{178.49 \text{ g Hf}}{1 \text{ mol Hf}} = 5.53 \text{ g Hf}$$

Response (e).

12-91   The number of mols of electrons is

$$16.5 \text{ h} \times \frac{3600 \text{ s}}{1 \text{ h}} \times \frac{1 \text{ C}}{\text{amp s}} \times 1.00 \text{ amp} \times \frac{1 \text{ mol e}^-}{96485 \text{ C}} = 0.616 \text{ mol e}^-$$

The mol Ce = $21.6 \text{ g Ce} \times \dfrac{1 \text{ mol Ce}}{140.12 \text{ g Ce}} = 0.154 \text{ mol Ce}$

Ratio $\dfrac{\text{mol Ce}}{\text{mol e}^-} = \dfrac{0.154}{0.616} = 0.25$; $\dfrac{\text{mol e}^-}{\text{mol Ce}} = \dfrac{0.616}{0.154} = 4$

Therefore, for every mol Ce there are 4 mol e$^-$ and the cerium ion has an oxidation state of 4$^+$. The formula is $CeCl_4$, x = 1, y = 4.

12-93 $15 \text{ min} \times \dfrac{60 \text{ s}}{1 \text{ min}} \times \dfrac{1 \text{ C}}{1 \text{ amp s}} \times 2.50 \text{ amp} \times \dfrac{1 \text{ mol e}^-}{96485 \text{ C}} = 0.0233 \text{ mol e}^-$

mol Au = $1.53 \text{ g Au} \times \dfrac{1 \text{ mol Au}}{196.97 \text{ g Au}} = 7.77 \times 10^{-3} \text{ mol Au}$

$\dfrac{\text{mol e}^-}{\text{mol Au}} = \dfrac{0.0233}{0.00777} = 2.99 \implies 3$

The oxidation number of gold is (3+).

12-95   At the anode the following reaction will take place: $2F^- \text{ (aq)} \rightarrow F_2(g) + 2 \text{ e}^-$

At the cathode the following reaction will take place: $Ca^{2+} \text{ (aq)} + 2 \text{ e}^- \rightarrow Ca(s)$

12-97   Water has a more positive standard reduction potential than Na^+ ions. Therefore, water will be reduced at the cathode, forming $H_2(g)$ and OH^- ions. At the anode Cl^- ion is oxidized to Cl_2 gas. There are various reasons why Cl^- ion is preferentially oxidized to Cl_2 rather than H_2O to $O_2(g)$. The most compelling of these is the very high over-potential for the oxidation of water.

12-99   Pure water has so few moles of ions per liter that the necessary electric current cannot be passed through the solution. Addition of an ionic substance greatly improves the electrolytic conduction so that a reaction can be sustained.

12-101   The reaction associated with electrolysis of water which occurs at the cathode, $2H_2O(l) + 2e^- \rightarrow H_2(g) + 2OH^-(aq)$, generates a basic environment due to the production of hydroxide ions.

12-103   Aluminum metal cannot be prepared by the electrolysis of an aqueous solution of the Al^{3+} ion. The reduction potential required is so great that only hydrogen gas will be liberated.

12-105   (a) Zinc is the anode and chromium is the cathode.
(b) Electrons flow from the Zn electrode to the cathode.
(c) The cathode is positive and the anode is negative.
(d) anode: $Zn(s) \rightarrow Zn^{2+}(aq) + 2e^-$ cathode: $Cr^{3+}(aq) + 3e^- \rightarrow Cr(s)$
(e) $3 Zn(s) + 2 Cr^{3+}(aq) \rightarrow 3 Zn^{2+}(aq) + 2 Cr(s)$
(f) Since Cr has a negative reduction potential, the reaction with the SHE would still be spontaneous and electrons would flow from the hydrogen electrode to the chromium electrode. The reaction would be $3 H_2(g) + 2 Cr^{3+}(aq) \rightarrow 6H^+(aq) + 2 Cr(s)$.

12-107 The heroine can escape the villains by using her knowledge of electrochemistry to corrode the various metals found in the room. Steel is made chiefly of Fe. If the exposed Cu pipe was placed in contact with the steel bars, Fe would be oxidized and any Cu ions found on the pipe would be reduced. This process would cause the steel bars to become brittle and break easily. She could also corrode the steel bars by coating them with 6 M HCl. Since brass is made up of a mixture of Cu and Zn, the HCl could oxidize the Zn and cause the window hinges to corrode and break.

12-109 (a) and (b)

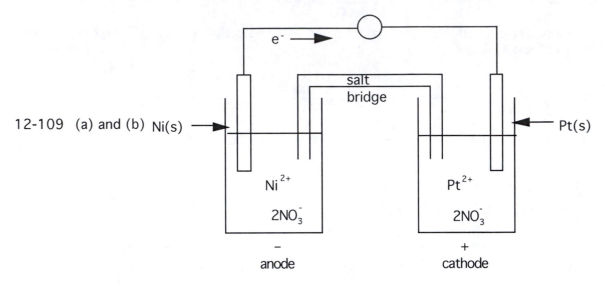

(c) No because the resulting $E_{cell}° = -0.21V$.
(d) The nickel will dissolve in the solution and platinum metal will form.

12-111 (a) Ag
        (b) $E_{cell}° = 1.56$ V
        (c) The Zn cell is the anode and the Ag cell is the cathode.
        (d) Electrons flow from Zn to Ag.
        (e) $2 Ag^+(aq) + Zn(s) \rightarrow 2 Ag(s) + Zn^{2+}(aq)$
        (f) at T=298K, $K = e^{\frac{nFE°}{RT}}$, where n=2,
             $K = e^{121.5} = 5.8 \times 10^{52}$

12-113 $3H_2C_2O_4(aq) + 2CrO_4{}^{2-}(aq) + 10H^+(aq) \rightarrow 6 CO_2(g) + 2Cr^{3+}(aq) + 8H_2O(l)$

$$40.0 \text{ mL CrO}_4{}^{2-} \times \frac{0.0250 \text{ mol CrO}_4{}^{2-}}{1000 \text{ mL}} \times \frac{3 \text{ mol H}_2C_2O_4}{2 \text{ mol CrO}_4{}^{2-}} \times \frac{1}{0.0100 \text{ L}} = 0.150 \text{ mol H}_2C_2O_4$$

12-115 $5Fe^{2+}(aq) + MnO_4{}^-(aq) + 8H^+(aq) \rightarrow Mn^{2+}(aq) + 4 H_2O (l) + 5Fe^{3+}(aq)$

$$3.2 \text{ mL MnO}_4{}^- \times \frac{3.0 \text{ mol MnO}_4{}^-}{1000 \text{ mL}} \times \frac{5 \text{ mol Fe}^{2+}}{1 \text{ mol MnO}_4{}^-} \times \frac{1 \text{ mol FeCl}_2}{1 \text{ mol Fe}^{2+}} \times \frac{126.753 \text{ g FeCl}_2}{1 \text{ mol FeCl}_2}$$

$= 6.1$ g FeCl$_2$

12A-1 This is an oxidation-reduction reaction
$3 P_4(s) + 10KClO_3(s) \rightarrow 3P_4O_{10}(s) + 10KCl(s)$

12A-3 (a)$5 Fe^{2+}(aq) + MnO_4^-(aq) + 8 H^+(aq) \rightarrow 5Fe^{3+}(aq) + Mn^{+2}(aq) + 4 H_2O(l)$
(b)$5 S_2O_3^{2-}(aq) + 8 MnO_4^-(aq) + 14 H^+(aq) \rightarrow 10 SO_4^{2-}(aq) + 8 Mn^{+2}(aq) + 7 H_2O(l)$
(c)$5 PbO_2(s) + 4 H^+(aq) + 2 Mn^{2+}(aq) \rightarrow 5 Pb^{2+}(aq) + 2 H_2O(l) + 2 MnO_4^-(aq)$
(d)$2 MnO_4^-(aq) + 6 H^+(aq) + 5 SO_2(g) + 2 H_2O(l) \rightarrow 2 Mn^{2+}(aq) + 5 H_2SO_4(aq)$

12A-5 (a)$H_2O_2(aq) \rightarrow H_2O(l) + O_2(g)$
oxidation half-reaction: $H_2O_2(aq) \rightarrow O_2(g)$
reduction half-reaction: $H_2O_2(aq) \rightarrow H_2O (l)$
For the oxidation half-reaction:
$H_2O_2(aq) \rightarrow O_2(g) + 2H^+(aq) + 2e^-$
For the reduction half-reaction:
$2e^- + 2H^+(aq) + H_2O_2(aq) \rightarrow 2H_2O(l)$
Overall reaction:
$2H_2O_2(aq) \rightarrow O_2(g) + 2H_2O(l)$

(b)$NO_2(g) + H_2O(l) \rightarrow HNO_3 (aq) + NO(g)$
oxidation half-reaction: $NO_2(g) \rightarrow HNO_3 (aq)$
reduction half-reaction: $NO_2(g) \rightarrow NO(g)$
For the oxidation half-reaction: $2(NO_2(g) + H_2O (l) \rightarrow HNO_3 (aq) + H^+ (aq) + e^-)$
For the reduction half-reaction: $2e^- + 2H^+ (aq) + NO_2(g) \rightarrow NO(g) + H_2O(l)$
Overall reaction: $3NO_2(g) + H_2O(l) \rightarrow 2HNO_3 (aq) + NO(g)$

12A-7 (a)$6HCl(aq) + 2HNO_3(aq) \rightarrow 2NO(g) + 3Cl_2(g) + 4H_2O (l)$
(b)$2HBr(aq) + H_2SO_4(aq) \rightarrow SO_2 (g) + Br_2(aq) + 2 H_2O(l)$
(c)$10HCl(aq) + 6H^+ (aq) + 2MnO_4^-(aq) \rightarrow 5Cl_2 (g) + 2Mn^{2+} (aq) + 8H_2O(l)$

12A-9 oxidation half-reaction: $H_2S (g) \rightarrow S_8(s)$
reduction half-reaction: $H_2O_2(aq) \rightarrow H_2O(l)$
For the oxidation half-reaction: $8H_2S (g) \rightarrow S_8(s) + 16H^+(aq) + 16e^-$
For the reduction half-reaction: $8(2e^- + 2H^+(aq) + H_2O_2(aq) \rightarrow 2H_2O(l))$
Overall reaction: $8H_2S(g) + 8H_2O_2(aq) \rightarrow S_8(s) + 16H_2O(l)$

12A-11 (a) For the oxidation half-reaction:

$$5(H_2O(l)+CH_3OH \text{ (aq)} \xrightarrow{H^+} CO_2(g) +6H^+(aq) +6e^-)$$

For the reduction half-reaction:

$$6(5e^-+8H^+(aq)+MnO_4^-(aq) \xrightarrow{H^+} Mn^{2+}(aq)+4H_2O(l))$$

Overall reaction:

$$5CH_3OH \text{ (aq)}+18H^+(aq)+6MnO_4^-(aq) \xrightarrow{H^+} 5CO_2(g)+6Mn^{2+}(aq)+ 19H_2O(l)$$

(b) For the oxidation half-reaction:

$$3(4OH^-(aq)+CH_3OH \text{ (aq)} \rightarrow HCO_2H(aq)+3H_2O(l)+4e^-)$$

For the reduction half-reaction:

$$4(3e^-+2H_2O(l) +MnO_4^-(aq) \rightarrow MnO_2(s) + 4OH^-(aq))$$

Overall reaction:

$$3CH_3OH \text{ (aq)} + 4MnO_4^-(aq) \rightarrow 3HCO_2H(aq) + H_2O(l) + 4MnO_2(s) + 4OH^-(aq)$$

12A-13 (a) $2SO_3^{2-}(aq) + O_2(g) \xrightarrow{H^+} 2SO_4^{2-}(aq)$

(b) $2SO_3^{2-}(aq) + O_2(g) \xrightarrow{OH^-} 2SO_4^{2-}(aq)$

12A-15 $2NH_3(aq) + ClO^- \text{ (aq)} \rightarrow N_2H_4 \text{ (aq)} + H_2O \text{ (l)} + Cl^-(aq)$

12A-17 $CuO(s) + 2 NH_3(g) \rightarrow 3 Cu(g) + 3H_2O(l) + N_2(g)$

12A-19 (a) CH_3COH is the oxidized species.

(b) CH_3COCH_3 is the oxidized species.

(c) Both have the same oxidation state.

(d) CO is the oxidized species.

Chapter 13
Chemical Thermodynamics

Note to student: Unless specifically directed by the problem, all thermodynamic values used in the solutions for this chapter are for atom combination rather than formation.

13-1   In referring to chemical reactions **spontaneous** refers to the fact that the reaction is favorable. Barring kinetic considerations, it will proceed from reactants to products.

13-3   Yes, the freezing of liquid water is spontaneous below 0°C but not spontaneous above 0°C.

13-5   Gaseous water has more entropy, because the gas has less constraints on its motion and more disorder in the system.

13-7   As warm water molecules come in contact with the penny, there is a transfer of energy to the penny. As this happens the average kinetic energy (temperature) of the penny increases and the average kinetic energy of the water decreases until they are equal. When the system has a uniform temperature the disorder, as measured by the distribution of energies (see chapter 6), is at its greatest.

13-9   If ΔS is negative, the system becomes more ordered. Response (d). A gaseous reactant forms a solid product.

13-11   Entropy increases as the disorder of the system increases. Response (c).

13-13   An entropically favored process is one in which ΔS_{sys} is positive. The process is not spontaneous if the ΔS_{univ} is negative as a result of the process.

13-15   All solutions have a concentration of 1.0 M.
        All gases have a partial pressure of 1.0 bar (0.9869 atm).

13-17   If the crystal is in perfect order and there is no inherent disorder and the entropy is zero. As the temperature rises, the motions of the particles within the crystal increase. The inherent disorder of these motions leads to an increase in entropy.

13-19   Gas has the highest entropy, because there is the least amount of structure and order in that state of matter. Solids generally have a definite structure. Liquids, while lacking definite shape, do not have as great a freedom of motion as gases.

13-21   If $\Delta S°$ is positive it means that entropy increased from reactants to products. So the products must have more entropy than the reactants. The opposite is true when $\Delta S°$ is negative.

13-23 $\Delta S° = 2034.6 + 6 \times 320.57 - 4 \times 1374 = -1538$ J/mol$_{rxn}$K
        Entropy is most likely not a driving force in the reaction.

13-25 $\Delta S° = 3 \times 116.972 - 2 \times 244.24 = -137.56$ J/mol$_{rxn}$K
        The negative $\Delta S°$ indicates that the products of the reaction are more ordered than the reactants.

13-27   The standard-state Gibbs energy change ($\Delta G°$) refers to a system under a unique set of standard conditions. ΔG refers to a system observed under any other of a wide variety of conditions.

13-29 $\Delta G° = \Delta H° - T\Delta S°$ for a reaction involving standard state species.
A negative value of $\Delta G°$ is associated with a spontaneous reaction. A spontaneous reaction will always result when $\Delta H° < 0$ and $\Delta S° > 0$.
Response (b).

13-31 $NH_3(g) \rightleftarrows NH_3(aq)$

$\Delta S° < 0$ because the gaseous phase is less ordered than the solution phase. $\Delta H° < 0$ because the hydrogen bonding potential should enhance the interactions in the water solution, but should not disrupt the water structure to any great extent. As the solution is heated, the ammonia gas is expelled. This happens because more $NH_3(aq)$ has enough energy to break the hydrogen bonds and become $NH_3(g)$.

13-33 $Fe_2O_3\ (s) + 2Al\ (s) \rightarrow Al_2O_3\ (s) + 2Fe\ (s)$

$\Delta H° = 2404.3 + 2 \times 326.4 - 3076.0 - 2 \times 416.3 = -851.5$ kJ/mol$_{rxn}$
$\Delta S° = 756.75 + 2 \times 136.21 - 761.33 - 2 \times 153.21 = -38.58$ J/mol$_{rxn}$K
The entropy change in this reaction is unfavorable, however the enthalpy change is so large that it is the driving force behind the reaction.

13-35 $2KMnO_4(s) + 5H_2O_2(aq) + 6H^+(aq) \rightarrow 2K^+(aq) + 2Mn^{2+}(aq) + 5O_2(g) + 8H_2O(l)$

$\Delta H°_{rxn} = 2 \times 2203.8 + 5 \times 1124.81 + 6 \times 217.65$
$\qquad\qquad -2 \times 341.62 - 2 \times 501.5 - 5 \times 498.340 - 8 \times 970.30 = -602.8$ kJ/mol$_{rxn}$
$\Delta S°_{rxn} = 2 \times 806.50 + 5 \times 407.6 + 6 \times 114.713$
$\qquad\qquad -2 \times 57.8 - 2 \times 247.3 - 5 \times 116.972 - 8 \times 320.57 = 579.7$ J/mol$_{rxn}$K

13-37 $PCl_5(g) + 4H_2O(l) \rightarrow H_3PO_4(aq) + 5HCl(g)$

$\Delta H° = 1297.9 + 4 \times 970.30 - 3241.7 - 5 \times 431.64 = -220.8$ kJ/mol$_{rxn}$
$\Delta S° = 624.60 + 4 \times 320.57 - 1374 - 5 \times 93.003 = 68$ J/mol$_{rxn}$K
All three, enthalpy change, entropy change, and Le Chatelier's principle, are driving forces in this reaction. $\Delta H°$ is negative, $\Delta S°$ is positive, and removal of HCl (g) will drive the reaction in the direction of product formation.

13-39 $3Fe(s) + 4H_2O(l) \rightarrow Fe_3O_4(s) + 4H_2(g)$

$\Delta G° = \Delta H° - T\Delta S°$
$\Delta H° = 3 \times 416.3 + 4 \times 970.30 - 3364.0 - 4 \times 435.30 = 24.9$ kJ/mol$_{rxn}$
$\Delta S° = 3 \times 153.21 + 4 \times 320.57 - 1039.3 - 4 \times 98.742 = 307.6$ J/mol$_{rxn}$K
$\Delta G° = 24.9$ kJ/mol$_{rxn}$ - T(0.3076 kJ/mol$_{rxn}$K)
At 298 K $\Delta G° = -66.8$ kJ/mol$_{rxn}$
At the higher temperatures of "white hot metal," the $T\Delta S°$ term in the expression for Gibbs energy change will dominate, giving a negative Gibbs energy change. The reaction will be spontaneous at the high temperature.

13-41   In 1 M acid, the reaction of Zn is $Zn(s) + 2H^+(aq) \rightleftarrows Zn^{2+}(aq) + H_2(g)$

$\Delta G^\circ = 95.145 + 2 \times 203.247 - 242.21 - 406.494 = -147.06 \text{ kJ/mol}_{rxn}$
This reaction is spontaneous.

$\Delta H^\circ = 130.729 + 2 \times 217.65 - 284.62 - 435.30 = -153.89 \text{ kJ/mol}_{rxn}$
$\Delta S^\circ = -23.1 \text{ J/mol}_{rxn}K$

In water,
$Zn(s) + 2H_2O(l) \rightarrow Zn^{2+}(aq) + 2OH^-(aq) + H_2(g)$
$\Delta G^\circ = 95.145 + 2 \times 875.354 - 242.21 - 2 \times 592.222 - 406.494$
      $= 12.71 \text{ kJ/mol}_{rxn}$
This reaction is not spontaneous.
$\Delta H^\circ = 130.729 + 2 \times 970.30 - 284.62 - 2 \times 696.81 - 435.30 = -42.2 \text{ kJ/mol}_{rxn}$
$\Delta S^\circ = -184.4 \text{ J/mol}_{rxn}K$

The products of this reaction are less stable than the reactants as shown by ΔG°. Both reactions have favorable enthalpy changes but the reaction with water has such an unfavorable entropy change that the reaction does not proceed.

13-43 $SiH_4(s) \rightarrow Si(s) + 2H_2(g)$
$\Delta H^\circ = 1291.9 - 455.6 - 2 \times 435.30 = -34.3 \text{ kJ/mol}_{rxn}$
$\Delta S^\circ = 422.20 - 149.14 - 2 \times 98.742 = 75.58 \text{ J/mol}_{rxn}K$
$\Delta G^\circ = -34.3 \text{ kJ/mol}_{rxn} - T(0.07558 \text{ kJ/mol}_{rxn}K)$
The reaction is spontaneous. Since $\Delta S^\circ > 0$ increasing the temperature would make the ΔG° even more negative, favoring the products even more.

13-45   (a) ΔG° will become more positive as T increases since $\Delta S^\circ < 0$.
       (b) ΔG° will become more negative as T increases since $\Delta S^\circ > 0$.
       (c) ΔG° will become more negative as T increases since $\Delta S^\circ > 0$.
       (d) ΔG° will become more positive as T increases since $\Delta S^\circ < 0$.

13-47 ΔG° will increase linearly with increasing temperature.

13-49   For $H_2O(s) \rightarrow H_2O(l)$
      at   -10°C $\Delta G > 0$
          0°C $\Delta G = 0$
      +10°C $\Delta G < 0$

13-51   The entropy increase associated with transition of water to the vapor phase is positive. The enthalpy change is positive since boiling is an endothermic process.

13-53 $\Delta G^\circ = \Delta H^\circ - T\Delta S^\circ$
$\Delta G^\circ_{500} = -484 \text{ kJ/mol}_{rxn} - 500 \text{ K}(-89 \text{ J/mol}_{rxn}K) = -440 \text{ kJ/mol}_{rxn}$
$\Delta G^\circ_{500} = -484 \text{ kJ/mol}_{rxn} - 1000 \text{ K}(-89 \text{ J/mol}_{rxn}K) = -395 \text{ kJ/mol}_{rxn}$
The calculated ΔG° values are more negative than the correct values. This indicates that either the ΔH° has a smaller magnitude at the higher temperatures or the ΔS° has a larger magnitude at the higher temperatures.

13-55 (a) $\Delta G° = -1882.25 - (-1040.156) - 2 \times (-406.494) = -29.11$ kJ/mol$_{rxn}$

(b) $\Delta G° = 2 \times (-532.0) - 3 \times (-463.462) = 326.4$ kJ/mol$_{rxn}$

(c) $\Delta G° = -980.1 + (-1529.078) - (-2639.5) = 130.3$ kJ/mol$_{rxn}$

13-57 From the "ideal gas relationship" the pressure is directly related to the mol/L concentration.

$PV = nRT$ $P = (n/V)RT$ where RT is a constant at a given temperature.

13-59 $K_p = K_c(RT)^{\Delta n}$

$\Delta n = 3 - 2 = 1$

$K_p = K_cRT$. Response (c).

13-61

	$COCl_2(g)$	$\rightleftarrows$	$CO(g)$	$+$	$Cl_2(g)$	$K_p = 3.2 \times 10^{-3}$
initial	0.124		0		0	
change	$-\Delta p$		Δp		Δp	
equilibrium	$0.124 - \Delta p$		Δp		Δp	

$$\frac{\Delta p^2}{0.124 - \Delta p} = 3.2 \times 10^{-3}$$

Assume $\Delta p \ll 0.024$

$\Delta p^2 = (0.124)(3.2 \times 10^{-3})$

$\Delta p = 1.99 \times 10^{-2}$

Check, $\dfrac{1.99 \times 10^{-2}}{0.124} \times 100\% = 16\%$, the assumption is not valid.

Solve using the quadratic equation.

$$\frac{\Delta p^2}{0.124 - \Delta p} = 3.2 \times 10^{-3}$$

$\Delta p^2 = (3.2 \times 10^{-3})(0.124 - \Delta p)$

$\Delta p^2 = 3.96 \times 10^{-4} - 3.2 \times 10^{-3}\Delta p$

$\Delta p^2 + 3.2 \times 10^{-3}\Delta p - 3.96 \times 10^{-4} = 0$

$\Delta p = 1.8 \times 10^{-2}$

$P_{COCl_2} = 0.11$ atm, $P_{CO} = P_{Cl_2} = 1.8 \times 10^{-2}$ atm

13-63

	2 $SO_3(g)$	$\rightleftarrows$	$O_2(g)$	+	2 $SO_2(g)$	$K_p = 1.5 \times 10^{-5}$
initial	0.490		0		0	
change	$-2\Delta p$		Δp		$2\Delta p$	
equilibrium	$0.490 - 2\Delta p$		Δp		$2\Delta p$	

$$K_p = \frac{P_{SO_2}^2 P_{O_2}}{P_{SO_3}^2} = \frac{(2\Delta p)^2 \Delta p}{(0.490 - 2\Delta p)^2} = 1.5 \times 10^{-5}$$

K_p is small and assume Δp is small

$$K_p \approx \frac{4\Delta p^3}{(0.490)^2} = 1.5 \times 10^{-5}$$

$\Delta p^3 = 9.0 \times 10^{-7}$

$\Delta p = 9.7 \times 10^{-3}$

$P_{SO_3} = 0.490 - 2(9.7 \times 10^{-3}) = 0.471$ atm

$P_{SO_2} = 2(9.7 \times 10^{-3}) = 1.9 \times 10^{-2}$ atm

$P_{O_2} = 9.7 \times 10^{-3}$ atm

Check, $K_p = \dfrac{(1.9 \times 10^{-2})^2(9.7 \times 10^{-3})}{(0.471)^2} = 1.6 \times 10^{-5}$

13-65

$$K_p = \frac{P_{NH_3}^4 P_{O_2}^7}{P_{NO_2}^4 P_{H_2O}^6} = 1.8 \times 10^{-28}$$

$$Q_p = \frac{(0.50)^4(0.50)^7}{(0.50)^4(0.50)^6} = \frac{(0.50)^{11}}{(0.50)^{10}} = 0.5 > K_p$$

The reaction will go to the left.

13-67   In this problem $K_p = K_c$

	$N_2(g)$	+	$O_2(g)$	$\rightleftarrows$	2 NO(g)	$K_p = 4.3 \times 10^{-9}$
initial	0.40		0.60		0	
change	$-\Delta c$		$-\Delta c$		$+2\Delta c$	
equilibrium	$0.40 - \Delta c$		$0.60 - \Delta c$		$2\Delta c$	

$$\frac{(2\Delta c)^2}{(0.40 - \Delta c)(0.60 - 3\Delta c)} = 4.3 \times 10^{-9} \text{ at } 700°C$$

Assume $\Delta c \ll 0.40$

$$\frac{4\Delta c^2}{(0.40)(0.60)} = 4.3 \times 10^{-9} \text{ at } 700°C$$

$\Delta c = 1.6 \times 10^{-5}$, the assumption is valid.

$[N_2]_{eq} = 0.40$ M

$[O_2]_{eq} = 0.60$ M

$[NO]_{eq} = 3.2 \times 10^{-5}$ M

A decrease in volume leads to a decrease in pressure. However, $\Delta n = 0$ so changing the volume has no effect.

13-69   (a) The first reaction will shift to the left. The other two will shift to the right.
(b) The ΔH for the first and third reactions is positive so and increase in temperature will shift the reaction to the right. The ΔH for the second reaction is negative so an increase in temperature will shift the reaction to the left.
(c) The addition of an inert gas at constant volume will have no effect on the reactions.

13-71   The larger the magnitude of $\Delta G°$ the farther the standard state conditions are from equilibrium.

13-73   A reaction system is at equilibrium when the Gibbs energy change, ΔG, is zero. Response (a).

13-75   Assume that the values of $\Delta H°$ and $\Delta S°$ at high temperature are not very different than the values given at 25°C.

$$2CH_4(g) \rightleftarrows C_2H_6(g) + H_2(g)$$

The tabulated values for the reaction are

Compound	$\Delta H_{ac}(kJ/mol_{rxn})$	$\Delta S_{ac}(J/mol_{rxn}K)$
$CH_4(g)$	-1662.09	-430.684
$C_2H_6(g)$	-2823.94	-774.87
$H_2(g)$	-435.30	-98.742

$\Delta H° = 2$ x 1662.09 - 2823.94 -435.30 = 64.94 kJ/mol_{rxn}
$\Delta S° = 2$ x 430.684 - 774.87 -98.742 = -12.24 $J/mol_{rxn}K$
At 1000 K
$\Delta G° = 64.94$ x 10^3 J/mol_{rxn} - 1000 K (-12.24 $J/mol_{rxn}K$) = 7.72 x 10^4 J/mol_{rxn}
$\Delta G° = -RT \ln K$

$$\ln K = \frac{-\Delta G°}{RT} = \frac{-\left(7.72 \times 10^4 \text{ J}\right)}{\left(8.314 \frac{J}{K}\right)(1000 \text{ K})} = -9.28$$

$K = e^{-9.28} = 9.3$ x 10^{-5}

13-77   For the reaction: $HCl(g) \rightleftarrows H^+(aq) + Cl^-(aq)$

From Table 13.2 ΔG = -35.9 kJ/mol_{rxn}. The value of K is calculated to be 2 x 10^6.
For the reaction: $CH_3CO_2H(aq) \rightleftarrows H^+(aq) + CH_3CO_2^-(aq)$

$\Delta H° = 3288.06 - 217.65 -3070.66 = -0.25$ kJ/mol_{rxn}
$\Delta S° = 918.5 - 114.713 -895.8 = -92.0$ $J/mol_{rxn}K$
$\Delta G° = 3015.42 - 203.247 - 2785.03 = 27.14$ kJ/mol_{rxn}

$$\ln K = \frac{-\Delta G°}{RT} = \frac{-\left(27.14 \times 10^3 \text{ J}\right)}{\left(8.314 \frac{J}{K}\right)(298.15 \text{ K})} = -10.95$$

$K = e^{-10.95} = 1.8$ x 10^{-5}. This value agrees with the value of K_a for acetic acid given in the Appendix.

13-79 First we need to recognize that there are 2 electrons being transferred in this process.

$$\Delta G^\circ = -nFE^\circ = -2 \cdot 96485 \frac{C}{mol} \cdot 0.76V = -1.5 \times 10^2 \frac{J}{mol}$$

$$\ln K = \frac{-\Delta G^\circ}{RT} = \frac{-\left(-1.5 \times 10^5 \; J\right)}{\left(8.314 \frac{J}{K}\right)(298.15 \; K)} = 60.5$$

$$K = e^{60.5} = 1.9 \times 10^{26}$$

13-81 First we need to recognize that there are 2 electrons being transferred in this process.

$$\Delta G^\circ = -nFE^\circ = -RT\ln K$$

$$E^\circ = \frac{-RT\ln K}{-nF} = \frac{8.314 \frac{J}{mol \cdot K} \cdot 298.15K \cdot \ln\left(1.8 \times 10^{-19}\right)}{2 \cdot 96485 \frac{C}{mol}} = -0.55V$$

13-83 $CO_2(s) + H_2(g) \rightleftarrows CO(g) + H_2O(g)$

$\Delta H^\circ = 1608.531 + 435.30 - 1076.377 - 926.29 = 41.16 \; kJ/mol_{rxn}$

$\Delta S^\circ\ \ = 266.47 + 98.742 - 121.477 - 202.23 = 41.51 \; J/mol_{rxn}K$

$\Delta G^\circ = 1529.078 + 406.494 - 1040.156 - 866.797 = 28.62 \; kJ/mol_{rxn}$

$$\ln K = \frac{-\Delta G^\circ}{RT} = \frac{-\left(28.62 \times 10^3 \frac{J}{mol}\right)}{\left(8.314 \frac{J}{mol \; K}\right)(298.15 \; K)} = -11.55$$

$K = e^{-11.55} = 9.6 \times 10^{-6}$. ΔH° is positive for this reaction and as shown in Chapter 10, K should increase with temperature. ΔS° is positive for this reaction. Increasing temperature makes ΔG° more negative, but as shown in the text, the behavior of ΔG° cannot always be relied upon for predictions about the temperature dependence of K.

13-85 The equilibrium constant for a reaction becomes 1 when $\Delta H^\circ = T\Delta S^\circ$. Assume the standard state enthalpy and entropy changes are approximately those at the temperature desired.

$\Delta H^\circ = 1297.9 \ - 966.7 - 243.358 = 87.8 \; kJ/mol_{rxn}$

$\Delta S^\circ\ \ = 624.60 \ - 347.01 - 107.330 = 170.26 \; J/mol_{rxn}K$

$$T = \frac{\Delta H^\circ}{\Delta S^\circ} = \frac{87.9 \times 10^3 J/mol_{rxn}}{170.26 \; J/mol_{rxn}K} = 516 \; K$$

13-87 $2SO_3(g) \rightleftarrows 2SO_2(g) + O_2(g)$

$\Delta H° = 2 \times 1422.04 - 2 \times 1073.95 - 498.34 = 197.84$ kJ/mol$_{rxn}$

$\Delta S° = 2 \times 394.23 - 2 \times 241.71 - 116.972 = 188.07$ J/mol$_{rxn}$K

$\Delta G° = \Delta H° - T\Delta S° = 197.79$ kJ/mol$_{rxn}$ - T(188.07 J/mol$_{rxn}$K)

$\ln K = \dfrac{-\Delta G°}{RT}$ and $K = e^{-\frac{\Delta G°}{RT}}$

T(K)	$\Delta G°$(J/mol$_{rxm}$)	ln K	K
298	1.42×10^3	-57.2	1×10^{-25}
473	1.09×10^3	-27.7	9×10^{-13}
673	71.2×10^3	-12.7	3×10^{-6}
873	33.6×10^3	-4.63	1×10^{-2}

The equilibrium constant increases with increasing temperature. ΔH is positive for this reaction. Therefore, this trend is expected.

13-89   (a) $\Delta S° = 2 \times (188.25) - 2 \times (130684) - (205.138) = -90.01$ J/mol$_{rxn}$K

           $\Delta H° = 2 \times (-241.818) - 2 \times (0) - 0 = -483.636$ kJ/mol$_{rxn}$

           $\Delta G° = -4.8363 \times 10^5$ J/mol$_{rxn}$ - 298.15 (-90.01 J/mol$_{rxn}$K)=
                 -457.144 kJ/mol$_{rxn}$

           $\Delta G° = 2 \times (-228.572) = -457.144$ kJ/mol$_{rxn}$

   (b) $\Delta S° = 2 \times (72.13) - 2 \times (51.21) - (223.066) = -181.226$ J/mol$_{rxn}$K

           $\Delta H° = 2 \times (-411.1553) - 2 \times (0) - 0 = -822.306$ kJ/mol$_{rxn}$

           $\Delta G° = -8.22306 \times 10^5$ J/mol$_{rxn}$ - 298.15 (-181.226 J/mol$_{rxn}$K)=
                 -768.476 kJ/mol$_{rxn}$

           $\Delta G° = 2 \times (-384.138) = -768.476$ kJ/mol$_{rxn}$

   (c) $\Delta S° = 2 \times (192.45) - 3 \times (130.684) - (191.61) = -198.762$ J/mol$_{rxn}$K

           $\Delta H° = 2 \times (-46.11) - 3 \times (0) - 0 = -92.22$ kJ/mol$_{rxn}$

           $\Delta G° = -9.222 \times 10^4$ J/mol$_{rxn}$ - 298.15 (-198.762 J/mol$_{rxn}$K)=
                 -32.90 kJ/mol$_{rxn}$

           $\Delta G° = 2 \times (-16.45) = -32.90$ kJ/mol$_{rxn}$

In all three cases the $\Delta G°$'s do not change with the method of calculation and are in agreement with the predictions from question 30.

13-91 $NH_3(g) \rightarrow NH_3(l)$

$\Delta H° = -80.29 - (-46.11) = -34.18$ kJ/mol$_{rxn}$

$\Delta S° = 111.3 - 192.45 = -81.15$ J/mol$_{rxn}$K

These values are in agreement with the prediction of problem 31.

13-93 $\Delta G°_{rxn}(HAc) = 2.71 \times 10^4$, $\Delta G°_{rxn}(HCl) = -3.59 \times 10^4$ Computer calculations were used to generate this table and significant figures were ignored.

Qc	ln Qc	ΔGrxn(HAc)	ΔGrxn(HCl)
1.00E-10	-23.026	-29948.204	-92948.204
1.00E-09	-20.723	-24243.383	-87243.383
1.00E-08	-18.421	-18538.563	-81538.563
1.00E-07	-16.118	-12833.742	-75833.742
1.00E-06	-13.816	-7128.9221	-70128.922
1.00E-05	-11.513	-1424.1018	-64424.102
1.00E-04	-9.21	4280.71858	-58719.281
1.00E-03	-6.908	9985.53894	-53014.461
1.00E-02	-4.605	15690.3593	-47309.641
1.00E-01	-2.303	21395.1796	-41604.82
1.00E+00	0	27100	-35900
1.00E+01	2.303	32804.8204	-30195.18
1.00E+02	4.605	38509.6407	-24490.359
1.00E+03	6.908	44214.4611	-18785.539
1.00E+04	9.21	49919.2814	-13080.719
1.00E+05	11.513	55624.1018	-7375.8982
1.00E+06	13.816	61328.9221	-1671.0779
1.00E+07	16.118	67033.7425	4033.74248
1.00E+08	18.421	72738.5628	9738.56283
1.00E+09	20.723	78443.3832	15443.3832
1.00E+10	23.026	84148.2035	21148.2035

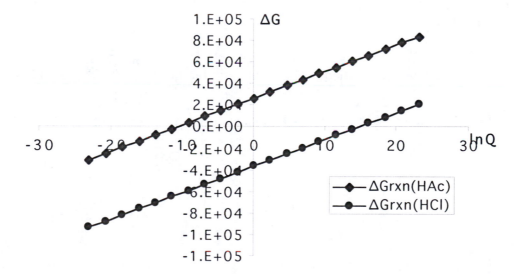

The graph shows that when ln Q_c = 0, $\Delta G_{rxn} = \Delta G°_{rxn}$. When ΔG_{rxn} = 0, the system is at equilibrium, and the point of intersection with the x axis is the ln K for the reaction.

13-95 $2 \, NO(g) + O_2(g) \rightleftharpoons 2 \, NO_2(g)$

(a) $\Delta H^\circ = 2 \times (-937.86) - 2 \times (-631.62) - (-498.340) = -114.74 \, kJ/mol_{rxn}$

(b) $\Delta S^\circ = 2 \times (-235.35) - 2 \times (-103.592) - (-116.972) = -146.54 \, J/mol_{rxn} \, K$

(c) $\Delta G^\circ = -114.74 \, kJ/mol_{rxn} - 298(-146.54 \, J/mol_{rxn}K) = -71.1 \, kJ/mol_{rxn}$

(d) At 298K the Gibbs energy for the reaction in the standard state is negative so it is a favorable reaction.

(e) $\ln K = \dfrac{-\Delta G^\circ}{RT} = \dfrac{-(-71.1 \times 10^3 \, J)}{\left(8.314 \dfrac{J}{K}\right)(298.15 \, K)} = 28.7$

$K = e^{28.7} = 2.86 \times 10^{12}$.

(f) $Q_p = \dfrac{P^2_{NO_2}}{P^2_{NO} \cdot P_{O_2}} = \dfrac{(1.0)^2}{(1.0)^2(1.0)} = 1.0$. Since Q < K the reaction will proceed to the right.

(g) If temperature is decreased the entropy term in ΔG° will decrease in magnitude, making the ΔG° more negative; this will increase the equilibrium constant.

13-97 Using the data from 13-95 and examining 13-95c in particular. Increasing the temperature will make the value of ΔG° more positive.

13-99 $H_2O(aq) \rightleftharpoons H^+(aq) + OH^-(aq)$

$\Delta H^\circ = -218 + (-696) - (-970) = 56 \, kJ/mol_{rxn}$

$\Delta S^\circ = -115 + (-286) - (-321) = 56 = -80 \, J/mol_{rxn} \, K$

$\Delta G^\circ = 56 \, kJ/mol_{rxn} - 298(-80 \, J/mol_{rxn}K) = 80 \, kJ/mol_{rxn}$

$CH_3COOH(aq) \rightleftharpoons H^+(aq) + CH_3COO^-(aq)$

$\Delta H^\circ = -218 + (-3071) - (-3288) = -1 \, kJ/mol_{rxn}$

$\Delta S^\circ = -115 + (-896) - (-919) = 56 = -92 \, J/mol_{rxn} \, K$

$\Delta G^\circ = -1 \, kJ/mol_{rxn} - 298(-92 \, J/mol_{rxn}K) = 26 \, kJ/mol_{rxn}$

Acetic acid will be stronger, even though both reactions have a positive ΔG°. The magnitude of the ΔG° for the first reaction is much greater, so it will be less likely to proceed than the second reaction.

13-101 Because the reaction quotient will change by a factor of $\dfrac{\left(n_{tot}\right)^2}{\left(n_{N_2}\right)^1}$ if N$_2$ is added at constant pressure this will actually lower Q_p and cause the reaction to move to the left.

Chapter 14
Kinetics

Note to student: The data plots given in the solutions to these problems were easily generated using Microsoft Excel©. The fitted lines were created by adding a trend line to the plot. It is a worthwhile exercise for you to learn how to generate these on your own.

14-1   **Thermodynamic control** of a reaction is when a reaction takes place the way the thermodynamics predict it should. That is, addition of magnesium metal to a strongly acidic solution will produce Mg^{2+} ions and hydrogen gas because the thermodynamics say that's what should happen. However, sometimes a reaction that is thermodynamically favored is kept from proceeding rapidly because it is under **kinetic control**. Such a reaction is the combustion of octane to produce carbon dioxide and water. While this is a thermodynamically favored reaction, it does not happen instantly when oxygen and octane are present in an automobile engine.

14-3   The rate is expressed in the form of a change in one of the reactants or products as a function of change in time.

14-5   As the reaction proceeds forward the phenolphthalein is consumed in the reaction, so as the concentration decreases (a negative change) we must show the rate as being positive (moving forward). Therefore we put a negative sign in front of the change in concentration of a reactant.

14-7   The instantaneous rate is given by rate $= -\dfrac{\Delta x}{\Delta t}$. Which is simply the slope of the curve of the concentration (x) at a given time (t).

14-9   For most reactions, the rate of reaction changes with time. The instantaneous rate of reaction is for a very short period of time. It can be measured more accurately than the changing rate of reaction, over a long period.

14-11   If the amount of N_2O_4 is measured in mol/liter, the units for the rate constant are $time^{-1}$.

14-13   If the amounts are measured in mol/liter, the units for the constant are $M^{-2}\ time^{-1}$.

14-15   3 mol F_2 are depleted for each 1 mol Cl_2. Thus the rate of depletion of F_2 is three times that of Cl_2. Response (e).

14-17   For each 2 mol NO_2 consumed, 1 mol N_2O_4 is formed.

$$\frac{d(N_2O_4)}{dt} = 0.0592 \text{ mol } NO_2 \times \frac{(N_2O_4)}{2(NO_2)} = 0.0296 \text{ Ms}^{-1}$$

14-19
$$-\frac{d(NH_3)}{dt} \times \frac{5 \text{ mol } O_2}{4 \text{ mol } NH_3} = -\frac{d(O_2)}{dt} \text{ . Therefore, } -\frac{d(NH_3)}{dt} = 0.800 \frac{d(H_2O)}{dt}$$

$$-\frac{d(O_2)}{dt} \times \frac{6 \text{mol} H_2O}{5 \text{mol} O_2} = \frac{d(H_2O)}{dt} \text{ . Therefore, } -\frac{d(O_2)}{dt} = 0.833 \frac{d(H_2O)}{dt}$$

131

14-21 $\dfrac{d(I_2)}{dt} = -\dfrac{d(MnO_4^-)}{dt} \times \dfrac{5\,mol\,I_2}{2\,mol\,MnO_{4^-}} = -(-4.56 \times 10^{-3}\ M\ s^{-1}) \times 2.50 = 0.0114\ M\ s^{-1}$

14-23 For simple gas phase reactions, the rate law for the reaction is more likely to reflect the observed stoichiometry of the reaction.

14-25 Unimolecular reactions depend upon the transformation of a single molecule independent of any other molecules present. Bimolecular reactions have rates that depend upon the transformation of two molecules, even if these two molecules happen to be the same compound.

14-27 A **rate-limiting step** is the step in a multi-step kinetic mechanism that is the slowest and acts as the "bottleneck" in the reaction.

14-29 If a reaction doesn't take place in a single step one must know all the simple steps (mechanism) of the reaction and which ones are slow (rate-limiting) to be able to write the rate law.

14-31 The first step and the slow, rate-determining step are one and the same. Therefore, Rate = $k(NO)^2(H_2)$

14-33 Response (a) fits the observed third order rate expression.
Response (b) describes a mechanism that requires a second-order rate equation.
Response (c) with a fast pre-rate determining step fits the stated observations. For the first step, $K_c = \dfrac{[N_2O_2]}{[NO]^2}$. For the second step Rate = $k(N_2O_2)(NO)$ or Rate = $kK(NO^2)(NO)$

Thus, responses (a) and (c) fit.

14-35 From the slow step (step 2): Rate = $k\,(NO_2)(NO_3)$
From step 1: $K_{eq} = \dfrac{[NO_2][NO_3]}{[N_2O_5]}$
$[NO_2][NO_3] = K_{eq}\,[N_2O_5]$
Substituting into the rate expression, Rate = $k\,K_{eq}\,(N_2O_5)$

14-37 For a one-step reaction, the equilibrium state can be identified with the equality of the rates for the forward and the reverse reactions.
[reactants] k_f = rate forward = rate backward = k_r [products]
k_f/k_r = [products]/[reactants] = K_c. The ratio of the forward to the reverse rate constants for a one-step reaction can be identified with the definition of the equilibrium constant.

14-39 Thinking about collision theory, decreasing the temperature will decrease the number of collisions particles will undergo, and therefore decrease the zero order rate constant.

14-41 Since the rate is not dependent upon the concentrations of the reactants, changing the concentrations will not change the rate.

14-43   Rate = k $(NO)^2(Cl_2)$

$$k = \frac{0.117\,Ms^{-1}}{(0.10M)^2(0.10M)} = 1.2 \times 10^2\ M^{-2}s^{-1}$$

Rate = $(1.2 \times 10^2\ M^{-2}s^{-1})\ (0.50\ M)^2\ (0.50\ M) = 15\ Ms^{-1}$

14-45   Rate = k $(NO)^2\ (O_2)$

$$k = \frac{0.355\,mmHg/s}{(100\,mmHg)^2(100\,mmHg)} = 3.55 \times 10^{-7}\ mmHg^{-2}s^{-1}$$

$$Rate = \frac{3.55 \times 10^{-7}}{mmHg^2 s}(250mmHg)^2(250\ mmHg) = 5.55\ mmHg\ s^{-1}$$

14-47   Rate = k $(CH_3I)(OH^-)$

$$k = \frac{8.78 \times 10^{-6}\,Ms^{-1}}{(1.35M)(0.10M)} = 6.5 \times 10^{-5}\ M^{-1}s^{-1}$$

Rate = $6.50 \times 10^{-5}\ M^{-1}s^{-1}\ (0.10\ M)(0.050\ M) = 3.3 \times 10^{-7}\ M\ s^{-1}$

14-49   This is the integrated rate law for reactions that are first-order in a single reactant. Unimolecular dissociation and nuclear decay reactions are first-order reactions. Problems where one might want to determine the amount of material left after a long period of time are best solved with the integrated rate law.

14-51   (a) For a second-order reaction, the integrated form of the rate law is

$$\frac{1}{(x)} - \frac{1}{(x)_o} = kt$$

$$\frac{\frac{1}{0.25M} - \frac{1}{0.50M}}{0.0050s} = k$$

k = $4.0 \times 10^2\ M^{-1}s^{-1}$

(b) $X_o = 0.50$ M and X = 0.15 M, $\frac{1}{(x)} - \frac{1}{(x)_o} = kt$,

$$\frac{\frac{1}{0.15M} - \frac{1}{0.50M}}{400\,M^{-1}s^{-1}} = t,\ t = 0.012\ sec$$

14-53   For first-order

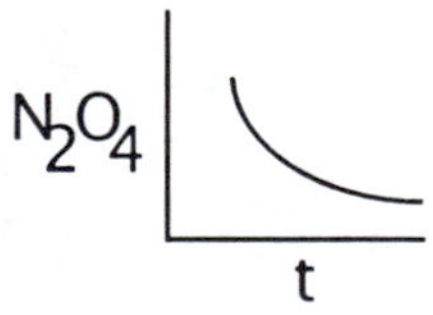

N_2O_4

14-55   For a second-order reaction

$$t_{1/2} = \frac{1}{k(x)_o}$$

Doubling $(x)_o$ would decrease the half-life by a factor of 2.

14-57 $t_{1/2} = \dfrac{0.693}{k} = \dfrac{0.693}{5.81 \times 10^{-11} \text{yr}^{-1}} = 1.19 \times 10^{10} \text{ yr}$

14-59 $\ln(0.795) = -kt$

$t_{1/2} = \dfrac{0.693}{k}$

$k = \dfrac{0.693}{t_{1/2}}$

$t = \dfrac{\ln(0.795)}{\dfrac{-0.693}{5730 \text{ yr}}} = 1.90 \times 10^3 \text{ yr}$

14-61 $\ln 0.057 = -kt$

$k = \dfrac{0.693}{t_{1/2}}$

$\dfrac{\ln(0.057)}{-0.693 / 5730 \text{ yr}} = 2.4 \times 10^4 \text{ yr}$

14-63 $\dfrac{\ln(0.903)}{-0.693 / 5730 \text{ yr}} = 844 \text{ yr}$; the beeswax did not belong to the Bronze Age.

14-65 A plot of the $\ln(N_2O)$ versus time gives a straight line of slope -k equal to the rate constant. Thus, the reaction is first-order. Rate$=k(N_2O)$

Reactant Conc. vs. Time

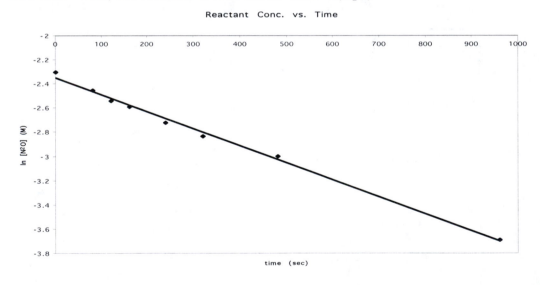

14-67   A plot of ln (BH_4^-) versus time gives a straight line indicating a first-order reaction.

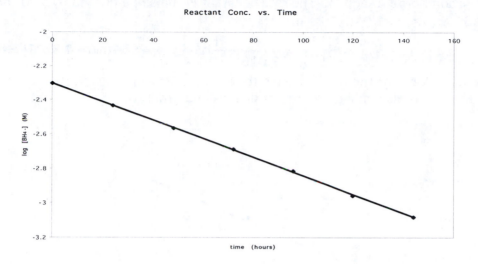

Reactant Conc. vs. Time

14-69   A plot of the ln ($Ni(CO)_4$) versus time is linear indicating that the reaction is first-order in $Ni(CO)_4$.

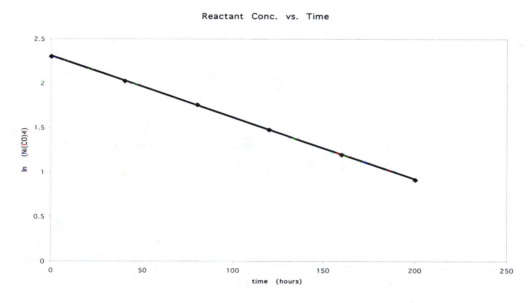

Reactant Conc. vs. Time

14-71   Because the reciprocal of concentration vs. time plot produces a linear set of data this must be a second-order reaction. The rate constant is the slope divided by 100 (note the change in the x axis) $k=1.016 \times 10^{-3}$ $min^{-1}M^{-1}$.

The half life is given as $t_{1/2}$ $= \dfrac{1}{k(X)_o} = \dfrac{1}{1.016 \times 10^{-3} \times 1.0} = 984\,min = 1.3 \times 10^2$ h.

It takes twice as long for the 0.5 M sample to reach 0.25M as it does for the 1.0 M to reach 0.5 M. This is also an indication that it is second-order.

Reactant Conc. vs. Time

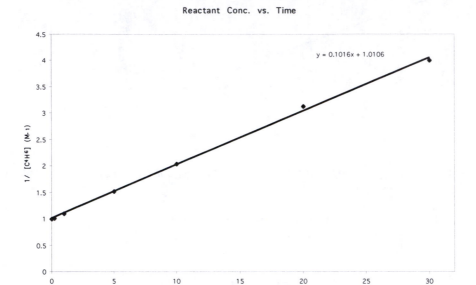

14-73   A plot of reactant concentration vs. time produces a linear plot. This means this is a zero order reaction and the slope of the plot is –k, so $k=0.015$ sec^{-1}.

Reactant Conc. vs. Time

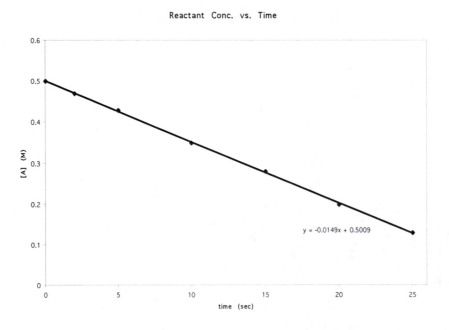

14-75　The reaction can be made pseudo-first-order in the presence of a relatively high hydroxide ion concentration. Alternatively, use of a buffer with sufficient capacity can also make the reaction appear pseudo first-order. The hydroxide ion concentration becomes essentially constant during the course of the reaction.

14-77　Since the rate doubles when the concentration of OH^- doubles, the reaction is first-order in OH^-. The overall order of the reaction is second-order. First-order in $Cr(NH_3)_5Cl^{2+}$ and first-order in OH^-.

$$\text{rate} = k \, (Cr(NH_3)_5\,Cl^{2+})(OH^-)$$

14-79　$t_{1/2} = \dfrac{0.693}{k} = \dfrac{0.693}{6.5 \times 10^{-9}\,s^{-1}} = 1.1 \times 10^8\,s$

14-81　That way one can study the order of the reaction with respect to each individual reactant.

14-83　The activation energy is the amount of energy that must be present in a reacting system for the system to be able to proceed to products. The higher the activation energy, the less likely an individual molecule will have that amount of energy and undergo reaction. The lower that likelihood, the slower the reaction.

14-85　The four properties are:
Catalysts increase the rate of reaction.
Catalysts are not consumed.
Catalysts do not affect ΔH or ΔS of a reaction.
Catalysts do not change the equilibrium constant.
A small amount of catalyst can affect the rate for a large amount of reactant. For example, the decomposition of H_2O_2 is catalyzed by iodide ion.

$$H_2O_2 + I^- \rightarrow H_2O + OI^-$$
$$OI^- + H_2O_2 \rightarrow H_2O + O_2 + I^-$$

Total　　$2H_2O_2 \rightarrow 2H_2O + O_2$

The I^- is available to react with another H_2O_2 molecule.

14-87　The enthalpy change of a reaction relates the difference in enthalpy between the reactant state and the product state. A catalyst will lower the energy required to get over the barrier from the reactant state, but has nothing to do with the enthalpy of the final products.

14-89　The catalyst drops the activation energy barrier by 30 kJ/mol$_{rxn}$ for the forward and for the reverse reaction. Therefore, the activation energy for the reverse reaction will be reduced to 155 kJ/mol$_{rxn}$.

14-91   Assume that the activation energy remains essentially constant for small changes in temperature.

$$\frac{k_1}{k_2} = e^{\frac{Ea}{R}\left(\frac{1}{T2}-\frac{1}{T1}\right)}$$

$$\ln\frac{k1}{k2} = \frac{E_a}{R}\left(\frac{1}{T_2}-\frac{1}{T_1}\right)$$

$$\ln\left(\frac{87.1}{1.53x103}\right) = \frac{Ea}{8.314}\left(\frac{1}{650}-\frac{1}{500}\right)$$

$$Ea = \frac{(-2.87)(8.314)}{-4.62x10^{-4}} = 5.16 \times 10^4 \text{ J/mol}_{rxn} = 51.6 \text{ kJ/mol}_{rxn}$$

14-93   Assume E_a does not change significantly with small changes in temperature.

$$\ln\frac{k1}{k2} = \frac{E_a}{R}\left(\frac{1}{T_2}-\frac{1}{T_1}\right)$$

$$\ln\left(\frac{3.5x10^{-7}}{k_2}\right) = \frac{183x10^3 \text{ J / mol}_{rxn}}{8.314 \frac{\text{J}}{\text{mol}_{rxn}\text{k}}}\left(\frac{1}{780K}-\frac{1}{550K}\right)$$

$$\ln(3.5 \times 10^{-7}) - \ln k_2 = (2.20 \times 10^4)(-5.36 \times 10^{-4})$$

$$-15 - \ln k_2 = -12; \quad -\ln k_2 = 3; \quad k_2 = e^{-3} = 0.05 \text{ M}^{-1}\text{s}^{-1}$$

14-95   If the activation energy (Ea) is lowered then the natural logarithm of the rate constant will increase in the Arrhenius relation. This means that the rate constant will increase exponentially with a drop in activation energy. Since the rate is proportional to the rate constant, there will also be an exponential increase in the rate of reaction.

14-97   There is a limit to how fast the enzyme can react with substrate. At high concentrations of substrate, the enzyme is operating as fast as it can. The reaction cannot go any faster, even if more substrate were added. In this situation, the rate of reaction is independent of the substrate concentration and the reaction is zero-order in sucrose.

14-99   (a) IV, (b) I, (c) III, (d) III,
Since the products are more stable than the reactants in II, this reaction would have K>1.
II and IV appear to have similar activation energies and would proceed at comparable rates.

14-101   The rate increases as the temperature increases, but not linearly.

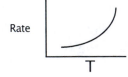

14-103   Addition of catalyst will affect the kinetics of the forward and reverse reactions, but not the thermodynamics of these reactions, therefore (b), (c), (d), and (e) will be affected by a catalyst. The others are thermodynamic functions.

14-105 (a) The half-life is 19.8 minutes.
(b) The third plot is correct since it is a first-order reaction. The data indicate that this reaction is following first-order kinetics since the half-life is independent of the concentration.
(It doesn't change when going from 0.1 M → 0.05 M as compared to 0.076 M → 0.038 M)

(c) $k = \dfrac{t_{1/2}}{0.693} = 0.035$ min^{-1}.

(d) The half-life would be longer at the lower temperature since there is less energy present in the system.
(e) This is an endothermic process, therefore the activation energy in the reverse direction will be less. It will be ΔH-Ea, which is 75 kJ/mol. Assuming that Z in the Arrhenius equation is the same for the forward and reverse reactions, a lower activation energy $k_{reverse}$ will be larger than $k_{forward}$ making the overall K <1.

14-107 Mechanism (b) is consistent with this rate law. The first step is the slow step and requires one molecule of A and one of B.

14-109 (a) The activation in the reverse direction is 226 kJ/mol + 134 kJ/mol = 400 kJ/mol.
(b) The collision may occur with the molecules oriented in the wrong way to permit a reaction. The molecules may not collide with enough energy to overcome the activation energy.
(c) The activated complex will be one in which the O-C-O bonds are in a line ready to form the CO_2 molecule.

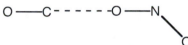

(d) Increasing the temperature typically increases the rate of a reaction since there are more collisions occurring and they are occurring with more energy.

14-111 (a) The mechanism implies a rate law of Rate= $k (CH_4)(Cl_2)^{1/2}$ because only one Cl of the two produced from the first reaction of Cl_2 is needed for the rate determining step and one methane molecule.
(b) The data is consistent with the rate law. When the methane concentration is doubled, the rate increases by a factor of two. When the (Cl_2) is doubled, however, the rate only increases by a factor of 1.41 which is the square root of two.

Chapter 15
Nuclear Chemistry

15-1 The alpha-particle is known to be equivalent to the He^{2+} ion. The beta-particle has been shown to be equivalent to the electron. Compared to other forms of electromagnetic radiation, gamma rays are very energetic.

15-3 The atomic number is the number of protons associated with the nucleus of an atom (and is also equal to the number of electrons for the neutral atom). The mass number for an atom represents the sum of the number of protons and neutrons in the nucleus of the atom. The term nuclide identifies a particular combination of protons and neutrons. The term nucleon identifies one of the components suggested to appear in the central core (or nucleus) of an atom.

15-5 38 electrons, 38 protons, and 52 neutrons.

15-7 The mass number is the sum of the number of protons and neutrons. During any of the three types of beta decay, the total number of neutrons and protons is constant. If a nuclide changes form during beta decay, the mass number must remain the same. Thus beta decay converts between isobars.

15-9 Gain or loss of neutrons interconverts isotopes, responses (e) and (f), and (g). Gain or loss of beta particles interconverts isobars, responses (a), (b), and (c). None of these interconvert isotones.

15-11 $^{99}_{42}Mo \rightarrow ^{0}_{-1}e + ^{99}_{43}Tc$

15-13 $^{62}_{29}Cu + ^{0}_{-1}e \rightarrow ^{62}_{28}Ni$

15-15 (a) $^{32}_{15}P \rightarrow ^{0}_{-1}e + ^{32}_{16}S$

(b) $^{11}_{6}C \rightarrow ^{0}_{+1}e + ^{11}_{5}B$

(c) $^{212}_{86}Rn \rightarrow ^{4}_{2}He + ^{208}_{84}Po$

(d) $^{125}_{54}Xe + ^{0}_{-1}e \rightarrow ^{125}_{53}I$

15-17 (a) $^{4}_{2}He$ $^{238}_{92}U + ^{4}_{2}He \rightarrow ^{239}_{94}Pu + 3\,^{1}_{0}n$

(b) $^{239}_{94}Pu$ $^{239}_{94}Pu + ^{4}_{2}He \rightarrow ^{242}_{96}Pu + ^{1}_{0}n$

(c) $4\,^{1}_{0}n$ $^{250}_{98}Cf + ^{11}_{5}B \rightarrow ^{257}_{103}Lr + 4\,^{1}_{0}n$

(d) $^{12}_{6}C$ $^{249}_{98}Cf + ^{12}_{6}C \rightarrow ^{257}_{104}Rf + 4\,^{1}_{0}n$

15-19 Electron capture and positron emission cause an increase in the number of neutrons and a decrease in the number of protons. The loss of two protons and two neutrons in alpha decay produces a smaller nuclide that is closer to the neutron:proton ratio of stable isotopes. Fission produces two smaller nuclides. Smaller nuclides are stable at a smaller neutron-to-proton ratio.

15-21 Neutron-poor nuclides have a mass number less than twice the atomic number of the element; (b) is neutron-poor, (a), (d), and (e) are neutron-rich, and (c) is on the line of stability.

15-23 Positron emission is associated with decay of a neutron-poor nuclide. Response (a).

15-25 When calculating binding energies, it is useful to first calculate the energy associated with 1 amu.

$$1 \text{ amu} = 1.6605655 \times 10^{-27} \text{ kg} \qquad c = 2.9979246 \times 10^8 \text{ m/s}$$
$$1 \text{ J} = \text{kg m}^2/\text{s}^2 \qquad 1 \text{ eV} = 1.6021892 \times 10^{-19} \text{ J}$$

$E = mc^2$

The energy for 1 amu is

$$E = \frac{1.6605655 \times 10^{-27} \text{ kg}}{1 \text{ amu}} \times \left(2.9979246 \times 10^8 \text{ m/s}\right)^2$$

$$\times \frac{1 \text{ eV}}{1.6021892 \times 10^{-19} \text{ J}} \times \frac{1 \text{ MeV}}{10^6 \text{ eV}} = \frac{931.5016 \text{ MeV}}{\text{amu}}$$

For $_3^6\text{Li}$

$$3 \text{ protons} \times \frac{1.00728 \text{ amu}}{\text{proton}} = 3.02184 \text{ amu}$$

$$3 \text{ neutrons} \times \frac{1.00867 \text{ amu}}{\text{neutron}} = 3.02601 \text{ amu}$$

$$3 \text{ electrons} \times \frac{0.0005486 \text{ amu}}{\text{electron}} = 0.001646 \text{ amu}$$

Total amu predicted = 6.04950 amu

mass defect = predicted amu - exact amu
mass defect = 6.04950 - 6.01512 = 0.03438 amu
This corresponds to a binding energy of

$$0.03438 \frac{\text{amu}}{\text{atom}} \times \frac{931.5016 \text{ MeV}}{\text{amu}} = 32.03 \frac{\text{MeV}}{\text{atom}}$$

Since there are six nucleons, the binding energy per nucleon is

$$32.03 \frac{\text{MeV}}{\text{atom}} \times \frac{1 \text{ atom}}{6 \text{ nucleons}} = 5.338 \frac{\text{MeV}}{\text{nucleon}}$$

15-27
$$7.570198 \frac{\text{MeV}}{\text{nucleon}} \times 238 \text{ nucleons} = 1.801707 \times 10^3 \frac{\text{MeV}}{\text{atom}}$$

$$1.801707 \times 10^3 \frac{\text{MeV}}{\text{atom}} \times \frac{1 \text{ amu}}{931.5016 \text{ MeV}} = 1.934196 \text{ amu}$$

Predicted amu - exact amu = mass defect = 1.934196 amu

Exact amu for $_{92}^{238}\text{U}$, 239.9861 amu -mass defect = 239.9861 amu -1.934196

The exact mass is 239.9861 - 1.9341 = 238.0520 amu

15-29 If we assume the energy to be associated with a loss of measured mass.
(10.0129 + 1.00867 - 4.00260 - 7.01600 amu) x 931.5016 MeV/amu =
2.8 MeV. The alpha particle would do more damage to the tumor tissue than other particles, but the damage would be short range.

15-31 In this case, we are looking for the sample to be cut in half, three times. From 1.000 gram to 0.500 gram in one half-life. From 0.500 gram to 0.250 gram is the second, and from 0.250 gram to 0.125 gram is the third. If three half-lives is 165 minutes, then one half-life is 55 minutes.

15-33　For a first-order reaction, $\ln\dfrac{0.20g}{1.0g} = -k(6.04\ yr)$

$k = 0.27\ yr^{-1}$ 　　　　$t_{1/2} = \dfrac{0.693}{k} = 2.6\ yr$

$\ln\left(\dfrac{0.075g}{1.00g}\right) = \dfrac{-0.27}{yr}\ t$

$\ln\left(\dfrac{0.075g}{1.00g}\right) = -2.6 = (-0.27\ yr^{-1})\ t$

$t = \dfrac{-2.59}{-0.27\ yr^{-1}} = 9.6\ yr$　to decay to 0.075 grams

15-35　$\ln\left(\dfrac{0.125\ g}{2.000\ g}\right) = -k\ (25d)$ 　　　$k = 0.11\ d^{-1}$

$t_{1/2} = \dfrac{0.693}{0.11\ d^{-1}} = 6.3\ d$

$1.875\ g\ ^{210}Po \times \dfrac{206\ g}{210\ g} = 1.84\ g\ ^{206}Pb$

15-37　Rate = kN

Rate = $1.00 \times 10^{-12}\ \dfrac{Ci}{L} \times 3.700 \times 10^{10}\ \dfrac{atoms}{s\text{-}Ci} = 0.0370\ \dfrac{atoms}{L\ s}$

$k = \dfrac{0.693}{t_{1/2}} = \dfrac{0.693}{3.823\ d \times 24\ \frac{h}{d} \times 3600\ \frac{s}{h}} = 2.10 \times 10^{-6}\ s^{-1}$

Rate = kN

$N = \dfrac{Rate}{K} = \dfrac{0.0370\ atoms}{L\ s \times 2.10 \times 10^{-6}\ s^{-1}} = 1.76 \times 10^{4}\ \dfrac{atoms}{L}$

15-39　(a)　Rate $= \dfrac{0.693 \times 0.00100g \times \left(6.022 \times 10^{23}\right)\ atom}{9.3\ min \times 60\ \frac{s}{min} \times 228\ g\ U \times \dfrac{3.700 \times 10^{10}\ \frac{atom}{s}}{Ci}} = 8.9 \times 10^{4}\ Ci$

(b)　Rate $= \dfrac{0.693 \times 0.00100g \times \left(6.022 \times 10^{23}\right)\ atom}{20.8\ d \times 24\frac{h}{d} \times 3600\frac{s}{h} \times 230g\ U \times \dfrac{3.700 \times 10^{10}\ \frac{atom}{s}}{Ci}} = 27.3\ Ci$

(c)

Rate $= \dfrac{0.693 \times 0.00100g \times \left(6.022 \times 10^{23}\right)\ atom}{(2.39 \times 10^{7})yr \times 365\frac{d}{yr} \times 24\frac{h}{d} \times 3600\frac{s}{h} \times 236g\ U \times \dfrac{3.700 \times 10^{10}\ \frac{atom}{s}}{Ci}} = 6.34 \times 10^{-8}\ Ci$

15-41 $\ln 0.057 = -kt$

$$k = \frac{0.693}{t_{1/2}}$$

$$\frac{\ln(0.057)}{-0.693/5730\,yr} = 2.4 \times 10^4\,yr$$

15-43 $\ln \frac{x}{x_o} = -kt, \quad t_{1/2} = \frac{0.693}{k}, \quad k = \frac{0.693}{t_{1/2}}$

$\ln \frac{x}{x_o} = \frac{-0.693}{5730y}(15520y) = -1.877; \quad \frac{x}{x_o} = \frac{^{14}C}{^{14}C_o} = e^{-1.877} = 0.153,\; 15.3\,\%$

15-45 $\dfrac{\ln(0.903)}{-0.693/5730\,yr} = 844\,yr$ The beeswax did not belong to the Bronze Age.

15-47 (a) Curies measure radioactivity in units that describe the amount of radiation given off. One Ci corresponds to 3.700×10^{10} disintegrations per second.
(b) Rads measure "radiation absorbed dose" which corresponds to the absorption of 10^{-5} Joules of energy per gram of body mass.
(c) Rems measure the "Roentgen equivalent man" which defines the toxicity of the radiation to biological systems. Rems = Rads x RBE.
(d) The Roentgen measures "the amount of radiation that will produce one electrostatic unit of ions per cubic centimeter volume."

15-49 $^{27}_{13}Al + {}^{4}_{2}He \rightarrow {}^{30}_{15}P + {}^{1}_{0}n$

$^{10}_{5}B + {}^{4}_{2}He \rightarrow {}^{13}_{7}N + {}^{1}_{0}n$

For ^{13}N, we should expect beta decay by electron capture or positron emission. Since ^{31}P is stable, a decay to increase the number of neutrons is probable. So we expect ^{30}P to undergo positron emission. Both are neutron poor.

15-51 $^{253}_{99}Es + {}^{4}_{2}He \rightarrow {}^{256}_{101}Md + {}^{1}_{0}n$

$^{238}_{92}U + {}^{19}_{9}F \rightarrow {}^{252}_{101}Md + 5\,{}^{1}_{0}n$

15-53 The tendency of nuclides to undergo fission or fusion reactions is closely related to the corresponding changes in binding energy per nucleon. For the lighter nuclides, fusion to produce larger nuclides leads to a greater binding energy per nucleon. Fission of the heavier nuclides does the same.

15-55 Collection and processing of materials containing the heavier nuclides required by fission reactors is costly. The by-products of the reaction generate wastes that are difficult to handle because of the radioactivity. Storage and containment become major problems. Containment seems to be the greatest problem associated with fusion reactors. The reaction is so highly exothermic, yet the reactants must remain in a plasma state. At this time usable energy is obtainable from a fission reactor, but science is not yet able to get more energy from fusion reactors than has to be put in to cause the reaction.

15-57 In the s-process, neutrons are captured one at a time to form a neutron-rich nuclide that has enough time to undergo alpha or beta decay before another neutron can be absorbed. This slow process cannot account for very heavy nuclides because the life times of the intermediate nuclei with atomic numbers between 83 and 90 are too short for this step-by-step absorption of neutrons to proceed. In the r-process, a number of neutrons are captured in rapid succession before there is time for alpha or beta decay to take place. Achieving an r-process reaction requires a very high neutron flux.

15-59 Since ^{48}Cr decay follows a first order process, a plot of ln mass ^{48}Cr versus time should produce a straight line whose slope is equal to the negative of the rate constant. Applying this procedure to the experimental data, we get
$k=5.0 \times 10^{-4}$ min^{-1}.
For a first order process, we know that the rate constant and half-life are related by

$$k = \frac{0.693}{t_{1/2}}$$

$$t_{1/2} = \frac{0.693}{k} = \frac{0.693}{5.0 \times 10^{-4} \text{ min}^{-1}} = 1.4 \times 10^3 \text{ min}$$

15-61 The detection limit is 0.10 in 15.3 so $\frac{x}{x_o} = \frac{0.10}{15.3}$, $\frac{\ln\frac{0.10}{15.3}}{-0.693 / 5730 \text{ yr}} = 4.2 \times 10^4$ yr

15-63 $E = h\upsilon$

$$= 6.626 \times 10^{-34} \text{Js} \times \left(\frac{3 \times 10^{17} \text{s}^{-1}}{\text{photon}}\right) \times \left(\frac{6.022 \times 10^{23}}{\text{mol}}\right) \times \left(\frac{1 \text{ kJ}}{1000 \text{ J}}\right)$$

$= 1 \times 10^5$ kJ/(mol photons)
This amount of energy could ionize approximately 100 mol of water.

Chapter 16
Organic Chemistry

16-1 Carbon forms covalent bonds with so many other elements because carbon exhibits an electronegativity which is too small for it to gain the needed electrons, but too large for carbon to lose the required number of electrons to form ionic bonds. Carbon, therefore, forms covalent bonds with a large number of other elements.

16-3 Saturated hydrocarbons contain as many hydrogen atoms as possible. An example is propane, $CH_3CH_2CH_3$. Unsaturated hydrocarbons have fewer hydrogen atoms than the corresponding alkane, and an example would be propene, $CH_2=CHCH_3$. In a straight chain hydrocarbon, the carbon atoms form a chain that runs from one end of the molecule to the other without carbon atoms branching off the main chain. Branched hydrocarbons do not form a single chain that runs from one end of the molecule to the other. Examples of both are given on the right.

$$CH_3 \diagup \overset{CH_2}{\diagdown} CH_2 \diagup \overset{CH_2}{\diagdown} CH_3$$
straight chain

$$H_3C \diagup \overset{\overset{\displaystyle CH_3}{|}}{CH} \diagdown CH_3$$
branched chain

16-5 For the alkane series of hydrocarbons, each carbon is bonded to four other atoms (either carbon or hydrogen). For n carbon atoms, there are 2n hydrogen atoms. Two additional hydrogens come from the terminal CH_3 groups. Alkanes, therefore, have the generic formula C_nH_{2n+2}. Two hydrogen atoms must be lost if a carbon chain is to be closed into a cyclic structure. The generic formula C_nH_{2n} describes a cycloalkane. An *alkene* has a carbon-carbon double bond. Two hydrogen atoms must be removed to allow for a C=C bond. The generic formula for an alkene with a single C=C is C_nH_{2n}, the same as that for a cycloalkane. Alkynes have a carbon-carbon triple bond. Four hydrogen atoms are displaced. The generic formula for an alkyne with one triple bond is C_nH_{2n-2}.

16-7 We can enumerate the isomers for heptane as:

1 for unbranched chain

C—C—C—C—C—C—C

2 for single branched chain

$$C—C—\overset{\overset{\displaystyle |}{C}}{\underset{\underset{\displaystyle C}{|}}{C}}—C—C—C \qquad C—\overset{\overset{\displaystyle |}{C}}{\underset{\underset{\displaystyle C}{|}}{C}}—C—C—C—C$$

4 for double branched chain

1 for triple branched chain

16-9 The molecular formula is $C_{12}H_{26}$. The name of the compound is 3,4-diethyl-3-methylheptane.

16-11 3,4,6-trimethyloctane

16-13

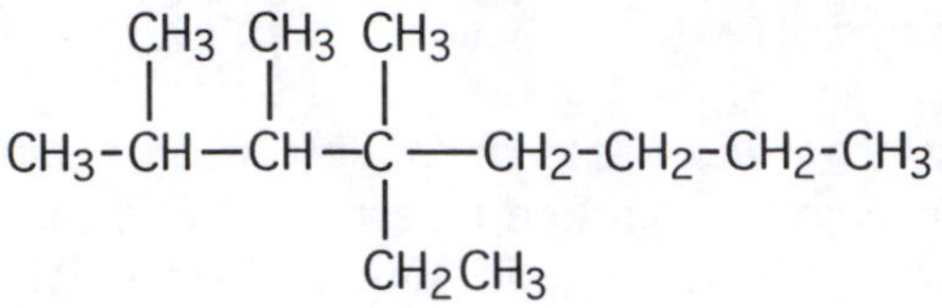

$$CH_3-CH-CH-C-CH_2-CH_2-CH_2-CH_3$$

with substituents CH_3, CH_3, CH_3 on the first three carbons and CH_2CH_3 below the central carbon.

16-15 The position of the C=C will move to the various unique positions. We will indicate the carbon skeleton.

C═C—C—C—C—C   1-hexene

C—C═C—C—C—C   2-hexene

C—C—C═C—C—C   3-hexene

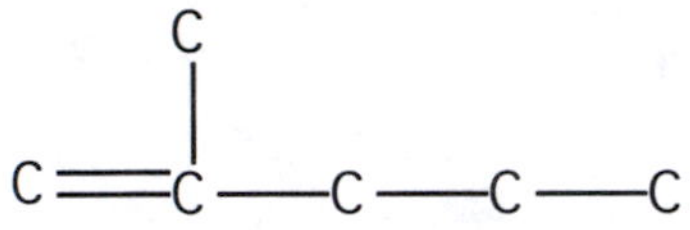

2-methyl-1-pentene 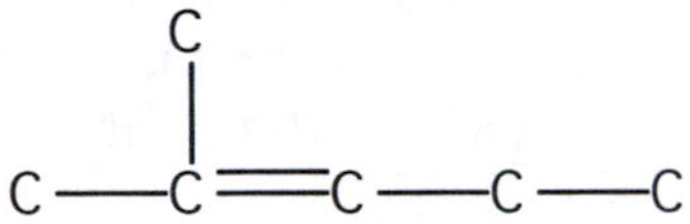   2-methyl-2-pentene 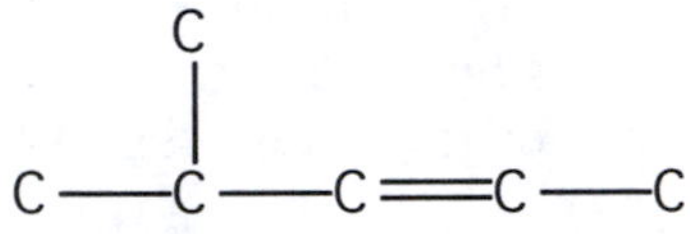   4-methyl-2-pentene

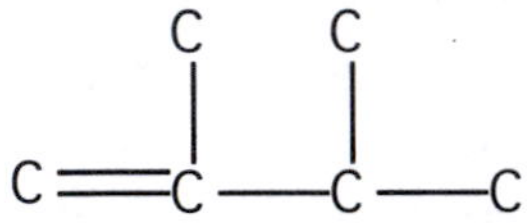

2,3-dimethyl-1-butene 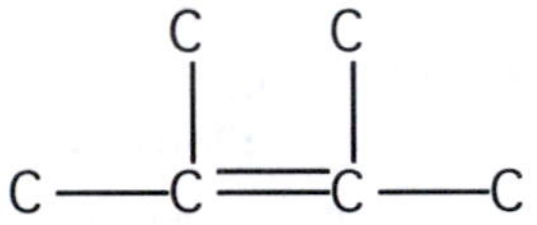   2,3-dimethyl-2-butene 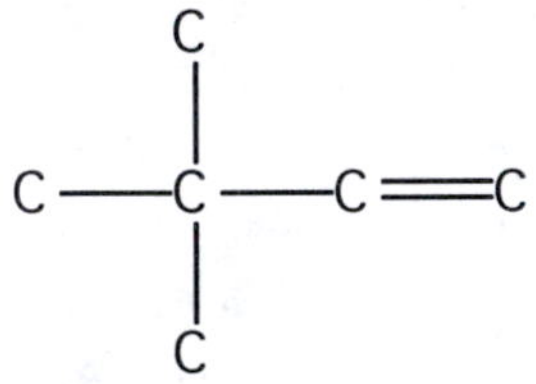   3,3-dimethyl-1-butene

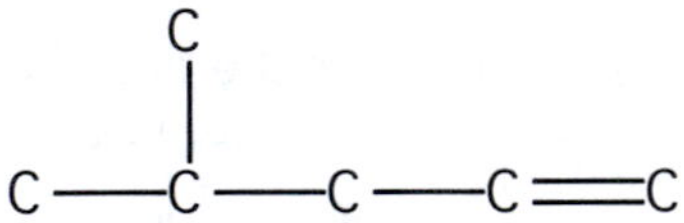

4-methyl-1-pentene 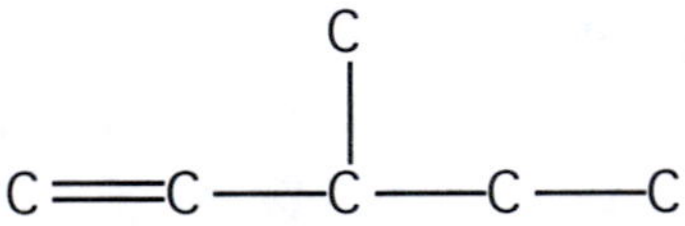   3-methyl-1-pentene 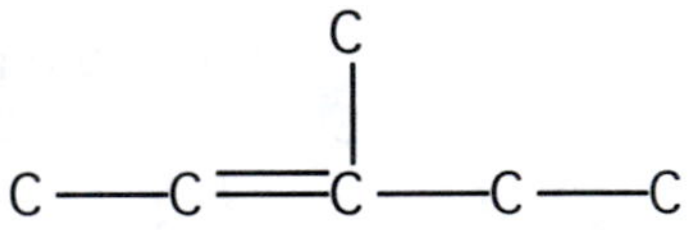   3-methyl-2-pentene

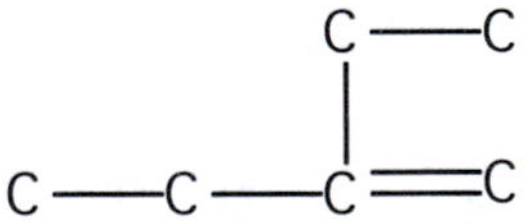

2-ethyl-1-butene

16-17 The principal functional group has the lowest number possible. Thus, if the carbons in this particular molecule were numbered starting at the other side of the molecule, the alkene functionality would start at carbon 2. Thus, the correct name should be 2-pentene.

16-19   Constitutional isomers are associated with changes in attachment of atoms. 1-butene and 2-butene are constitutional isomers in terms of the location of the C=C bond. Stereoisomers retain the same relative attachments, but the component parts of a molecule take up alternate positions relative to a "constraint" such as a C=C. The attachment of atoms around the double bond is the characteristic feature that can be used.

16-21   Response (d). Pentane has the molecular formula C_5H_{12} while all the other compounds have the molecular formula C_5H_{10}.

16-23   (a) C=C-C-C-C has no cis and trans isomers.
         (b) C-C=C-C-C has possible cis and trans isomers.
         (c) has no cis and trans isomers because the same group appears twice on a single carbon center of the C=C.
         (d) has no cis and trans isomers because the same group appears twice on a single carbon center of the C=C.
         (e) C≡C results in a linear geometry and there are no cis or trans isomers.

16-25

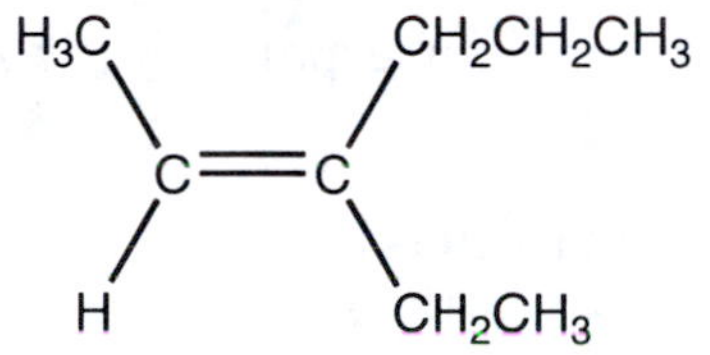

16-27

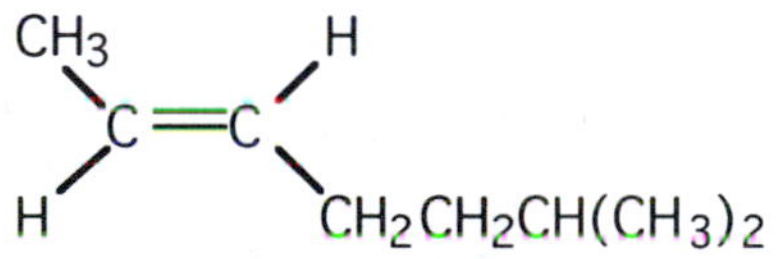

16-29

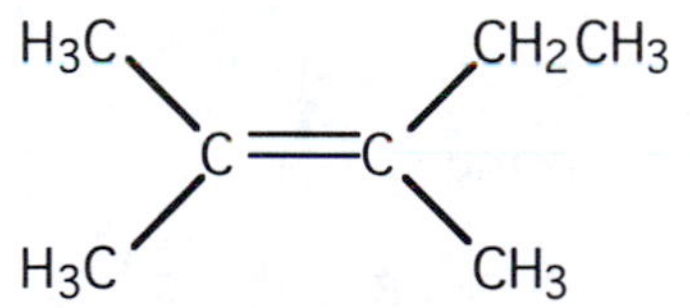

         Since two of the methyl (-CH3) substituents are found on the same carbon of the alkene functionality, this molecule cannot exhibit *cis/trans* isomerism.

16-31   The bond between the third and fourth carbons is Z.

16-33

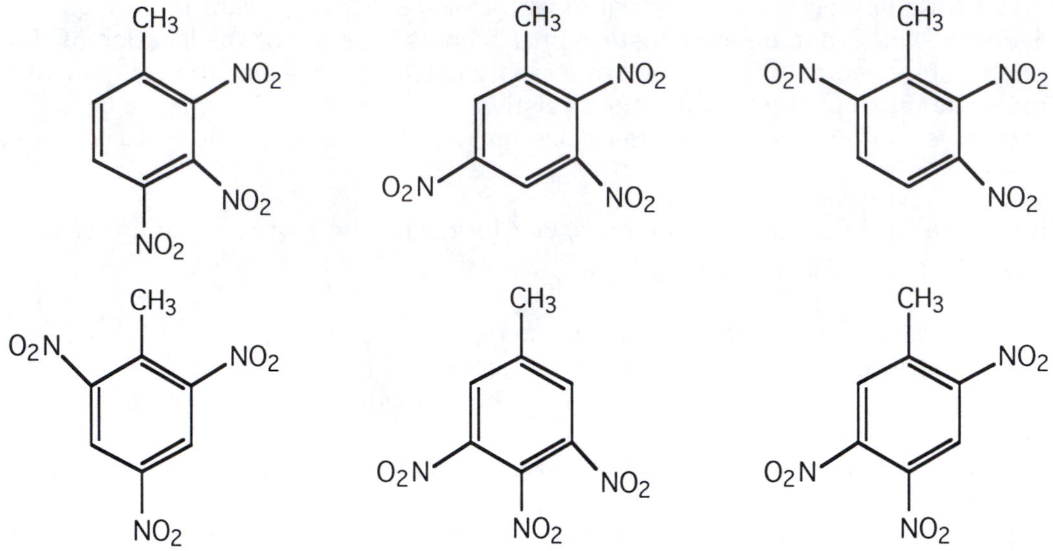

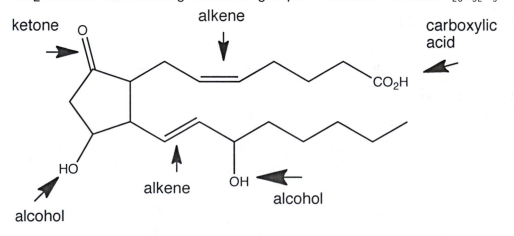

2,4,6-trinitrotoluene

16-35 Each carbon atom in benzene has three bonding domains with corresponding trigonal planar geometry.

16-37 The properties of the petroleum fractions are summarized in Table 16.3. As the average chain length increases, the boiling temperatures of the hydrocarbons increase.

16-39 Increasing the average length of the hydrocarbon chains will not increase the octane number of gasoline. Response (d)

16-41 (a) CH_3CH_2OH (b) CH_3CHO (c) CH_3NH_2 (d) CH_3CONH_2 (e) CH_3Br (f) $CH_2=CH_2$ (g) $HC\equiv CH$

16-43 (a)aldehyde (b)alcohol (c)ether (d)ketone

16-45 Molecular formula of cortisone is $C_{21}H_{28}O_5$. The functional groups include an alkene, two alcohol groups, and three ketone groups.

16-47 PGE_2 contains the following functional groups.   Molecular Formula: $C_{20}H_{32}O_5$

16-49

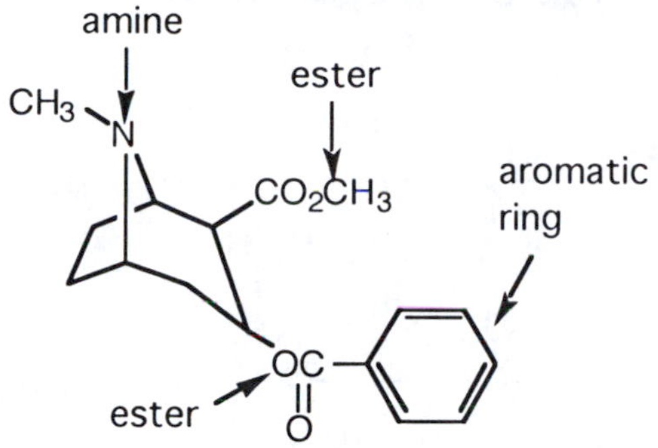

Cocaine structure labels: amine, CH₃, N, ester, CO₂CH₃, aromatic ring, ester, OC, O

CH_3 — amine

ester — CO_2CH_3

aromatic ring

ester — OC / O

Cocaine

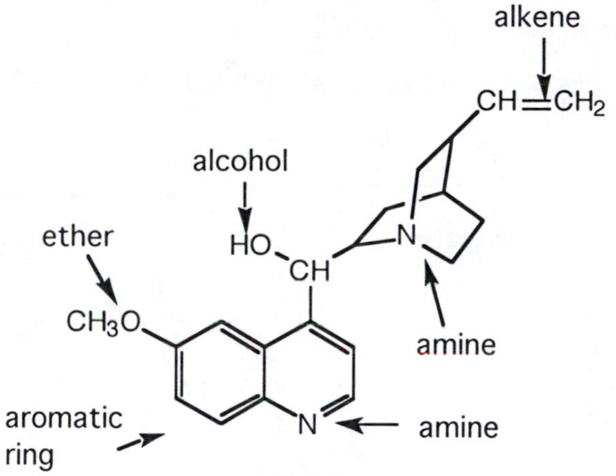

Quinine structure labels: alkene, $CH=CH_2$, alcohol, HO, CH, ether, CH_3O, amine, N, aromatic ring, amine, N, amine

Quinine

16-51 (a) Oxidation-reduction reaction   (b) Oxidation-reduction reaction
(c) Oxidation-reduction reaction   (d) Not an oxidation-reduction reaction
(e) Oxidation-reduction reaction

16-53 (a) A primary alcohol can be oxidized to form an aldehyde. (b) A secondary alcohol can be oxidized to form a ketone, but not an aldehyde. (c) Unless the ether is cleaved, no oxidation will occur. (d) A ketone is not further oxidized.

16-55 Reduction of an aldehyde generally yields the related primary alcohol without corresponding changes in the carbon chain. Response (b).

16-57 Oxidation of 2-methyl-3-pentanol, a secondary alcohol, will give 2-methyl-3-pentanone, a ketone.

$$CH_3-\underset{\underset{H}{|}}{\overset{\overset{CH_3}{|}}{C}}-\underset{\underset{H}{|}}{\overset{\overset{OH}{|}}{C}}-CH_2CH_3 \longrightarrow CH_3-\underset{\underset{H}{|}}{\overset{\overset{CH_3}{|}}{C}}-\overset{\overset{O}{||}}{C}-CH_2CH_3$$

16-59 Br_2 + energy → 2Br·   Initiation

CH_4 + Br· → CH_3· + HBr   Propagation

CH_3· + Br_2 → CH_3Br + Br· Termination

2Br· → Br_2

CH_3· + Br· → CH_3Br

16-61   The different intermediates result from abstraction of the various hydrogens.

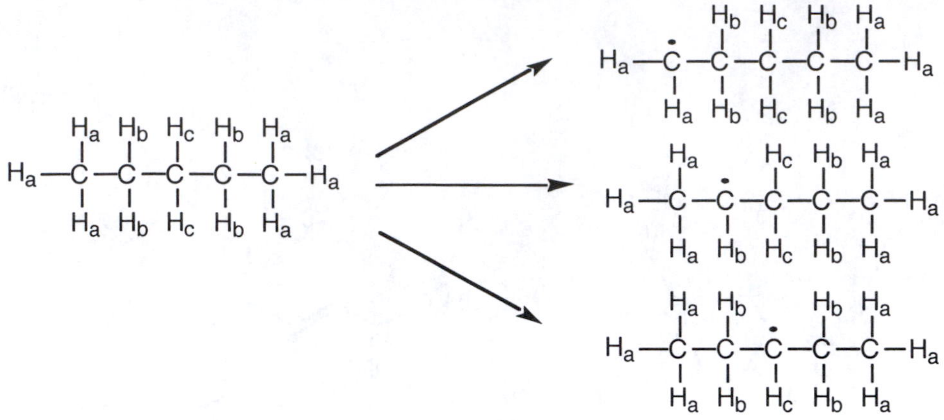

primary

secondary

secondary

16-63   Alcohols: $CH_3CH_2CH_2CH_2OH$, $(CH_3)_2CHCH_2OH$, $(CH_3)_3COH$, $CH_3CH_2CH(CH_3)OH$
Ethers: $CH_3CH_2OCH_2CH_3$, $CH_3CH_2CH_2OCH_3$, $(CH_3)_2CHOCH_3$

16-65   The "dehydration" of ethanol will yield diethyl ether.
$$2\ CH_3CH_2OH(l) + 2\ H^+(aq)\ \rightarrow\ 2\ CH_3CH_2OCH_2CH_3\ (l) + H_2O(l)$$

16-67

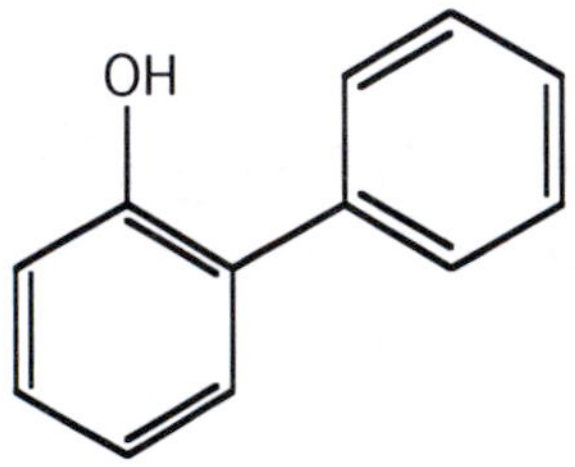

o-phenylphenol

16-69   The conjugate bases formed from alcohols acting as Brønsted acids will be of the general type $R-O^-$, where R represents the hydrocarbon part of the molecule.

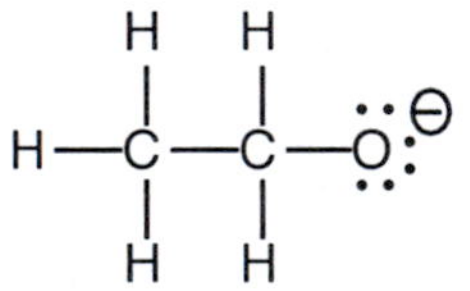

ethoxide 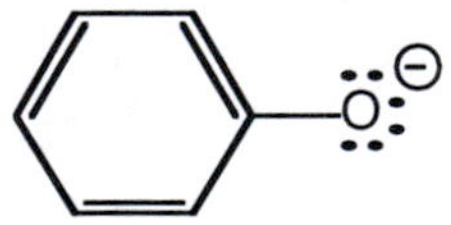   phenoxide 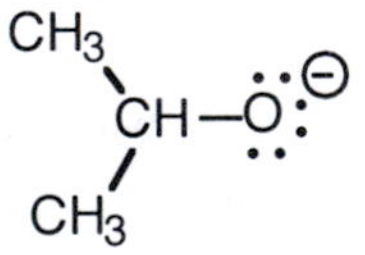   isopropyl alkoxide

16-71   A: 5-ethyl-3-octanol   B: 5-methyl-1-hepten-4-ol

16-73   (a) and (e) are electron deficient and will act as electrophiles toward the oxygen (electron rich) end of the carbonyl group. (b), (c) and (d) are electron rich and will act as nucleophiles toward the carbon (electron deficient) end of the carbonyl group.

16-75   Strong oxidizing agents cannot be used to convert a primary alcohol to an aldehyde, because in the presence of the excess oxidizing agent, any aldehyde formed would be oxidized to the corresponding carboxylic acid.

16-77   3,7-dimethyl-2,6-octadienal

16-79   The greater polarity of the carboxylic acid, the increased water solubility and the presence of an ionizable proton differentiate this functional group from the ester. Apparently man has inherited a sensitivity to the presence of decay products as signaled by the rather unpleasant odor associated with intermediate length carbon chain carboxylic acids. The odor of the corresponding esters signal that the source is ripe and ready for eating or use.

16-81   Nicotine exhibits both a tertiary amine of the traditional type and one associated with a delocalized valence electron ring structure (pyridine). Coniine is a cyclic secondary amine. Strychnine has a tertiary amine and a cyclic amide functional group. Lysergic acid dimethyl amide has a tertiary amine, secondary amine, and an amide functional group. The rest of the compounds exhibit tertiary amine functional groups.

16-83

caffeine

16-85   To be chiral, a structure must exhibit non-superimposable mirror images. For tetrahedral carbon centers, this means that all four bonded groups must be different. Only response (e) represents a structure that can be chiral.

16-87   Since it has a stereocenter at the substituted carbon next to the cyclic amine, coniine is optically active.

16-89   Vitamin C has two stereocenters.

16-91   Both isocitric and homocitric acid are chiral, since they possess carbon atoms with four different groups attached.

16-93   The relative configurations at the stereocenters can be used to distinguish the two. These two molecules are enantiomers.

R configuration   S configuration

16-95   R

16-97   Carboxylic acids give weakly acidic solutions in water. The acids of intermediate chain length have unpleasant odors. Esters generally have rather pleasant odors.

16-99   Primary alcohols will be oxidized by acid dichromate, tertiary alcohols will not.

16-101 Tartaric acid is chiral. It has two stereocenters. Malic acid is chiral.